T0255368

MENTAL CAPACITY IN RELATIONSHIP

Recent legal developments challenge how valid the concept of mental capacity is in determining whether individuals with impairments can make decisions about their care and treatment. Kong defends a concept of mental capacity but argues that such assessments must consider how relationships and dialogue can enable or disable the decision-making abilities of these individuals. This is thoroughly investigated using an interdisciplinary approach that combines philosophy and legal analysis of the law in England and Wales, the European Convention of Human Rights, and the UN Convention on the Rights of Persons with Disabilities.

By exploring key concepts underlying mental capacity, the investigation concludes that both primary relationships, and capacity assessments themselves must display key competencies to ensure that autonomy skills are promoted and encouraged. This ultimately provides scope for justifiable interventions into disabling relationships and articulates the dialogical practices that help better situate, interpret, and understand the choices and actions of individuals with impairments.

CAMILLIA KONG is a lecturer in philosophy at the University of Kent and research associate at the Ethox Centre, University of Oxford. Her research interests include philosophical issues surrounding mental disorder, mental capacity, and the ethics of psychiatry. She was awarded funding for research into this book from the British Academy.

CAMBRIDGE BIOETHICS AND LAW

This series of books was founded by Cambridge University Press with Alexander McCall Smith as its first editor in 2003. It focuses on the law's complex and troubled relationship with medicine across both the developed and the developing world. Since the early 1990s, we have seen in many countries increasing resort to the courts by dissatisfied patients and a growing use of the courts to attempt to resolve intractable ethical dilemmas. At the same time, legislatures across the world have struggled to address the questions posed by both the successes and the failures of modern medicine, while international organisations such as the WHO and UNESCO now regularly address issues of medical law.

It follows that we would expect ethical and policy questions to be integral to the analysis of the legal issues discussed in this series. The series responds to the high profile of medical law in universities, in legal and medical practice, as well as in public and political affairs. We seek to reflect the evidence that many major health-related policy debates in the UK, Europe and the international community involve a strong medical law dimension. With that in mind, we seek to address how legal analysis might have a trans-jurisdictional and international relevance. Organ retention, embryonic stem cell research, physician assisted suicide and the allocation of resources to fund health care are but a few examples among many. The emphasis of this series is thus on matters of public concern and/or practical significance. We look for books that could make a difference to the development of medical law and enhance the role of medico-legal debate in policy circles. That is not to say that we lack interest in the important theoretical dimensions of the subject, but we aim to ensure that theoretical debate is grounded in the realities of how the law does and should interact with medicine and health care.

MENTAL CAPACITY IN RELATIONSHIP

Decision-Making, Dialogue, and Autonomy

CAMILLIA KONG
University of Kent

CAMBRIDGE
UNIVERSITY PRESS

University Printing House, Cambridge CB2 8BS, United Kingdom

One Liberty Plaza, 20th Floor, New York, NY 10006, USA

477 Williamstown Road, Port Melbourne, VIC 3207, Australia

314-321, 3rd Floor, Plot 3, Splendor Forum, Jasola District Centre, New Delhi - 110025, India

79 Anson Road, #06-04/06, Singapore 079906

Cambridge University Press is part of the University of Cambridge.

It furthers the University's mission by disseminating knowledge in the pursuit of education, learning and research at the highest international levels of excellence.

www.cambridge.org
Information on this title: www.cambridge.org/9781316615706
DOI: 10.1017/9781316683088

First published 2017
First paperback edition 2018

A catalogue record for this publication is available from the British Library

ISBN 978-1-107-16400-0 Hardback
ISBN 978-1-316-61570-6 Paperback

For
Sophie and Nelson
and
Ava – always beloved

We are not signs,
we do not live in spite of
or because of our facts,
we live with them, around them, among,
like we live around rivers, my cane,
your warts, like we live among animals,
your heart, my brace, like we live
with each other.
– Jim Ferris, "Facts of Life"

CONTENTS

ACKNOWLEDGEMENTS

I would like to thank the British Academy for its generous funding and assistance under its Postdoctoral Fellowship Scheme, which enabled me to undertake the research for this book. I have been very fortunate to work at the Ethox Centre, which has been an overwhelmingly supportive, fertile research environment for the completion of the book. I also undertook a fruitful month of research hosted by the Department of Medical Humanities at the VU University Medical Center in Amsterdam – my gratitude to Guy Widdershoven and Suzanne Metslaar for their enriching intellectual exchange, helpful comments on chapters, and kind hospitality during my stay. Thanks to Richard Ashcroft for his suggestions during the book proposal phase, as well as to various individuals who read and provided constructive feedback on various chapters, including Gerben Meynen, Michael Parker, Charles Foster, Lucy Series, and three anonymous reviewers. I am especially indebted to Alex Ruck Keene, Tony Hope, Alice Obrecht, and Michael Dunn, whose detailed comments on the full draft of the book enriched the final product. Needless to say, any remaining flaws and mistakes reflect my own shortcomings. Finally, words cannot express my gratitude to Andrew de Jong Cleyndert, whose boundless encouragement, patience, and love has been my lodestar throughout.

LIST OF STATUTES

The Adult Guardianship and Co-decision-making Act, SS. 2000
European Convention on Human Rights
Health Care Consent Act, 1996, S.O. 1996
Mental Capacity Act 2005 in England and Wales
Mental Capacity Act Singapore
Substitute Decisions Act, 1992, S.O. 1992
United Nations Convention for the Rights of Persons with Disabilities

LIST OF CASES

A, B, & C v X & Y [2012] EWHC 2400 (COP)
Aintree University Hospitals NHS Foundation v James [2013] UKSC 67
Airedale NHS Trust v Bland [1993] 1 FLR 1026, 1035–1036
A Local Authority v A & Anor [2010] EWHC 1549 (Fam)
A Local Authority v DL & Ors [2010] EWHC 2675 (Fam)
A Local Authority v DL & Ors [2011] EWHC 1022 (Fam)
A Local Authority v E & Ors [2012] EWHC 1639 (COP)
A Local Authority v M & Ors [2014] EWCOP 33
A Local Authority v WMA & Ors [2013] EWHC 2580 (COP)
A Local Authority X v MM & Anor (No. 1) [2007] EWHC 2003 (Fam)
A Primary Care Trust v P & Ors [2009] EW Misc 10 (EWCOP)
Cheshire West and Chester Council v P & Or [2011] EWHC 1330 (Fam)
Cheshire West and *MIG and MEG* cases (reported sub nom *P v Cheshire West and Chester Council and another; P and Q v Surrey County Council* [2014] UKSC 19, [2014] 1 AC 896)
Glor v Switzerland App. No. 13444/04
ITW v Z & Ors [2009] EWHC 2525 (Fam)
KK v STCC [2012] EWHC 2136
LB Haringey v FG [2011] EWHC 3932 (COP)
LBL v RYJ & Anor [2010] EWHC 2665 (COP)
LBX v K, L, and M [2013] EWHC 3230 (Fam) (2013) MLO 148
London Borough of Redbridge v G, C, and F [2014] EWHC 485 (COP)
Loughlin v Singh & Ors [2013] EWHC 1641 (QB)
Malette v Shulman [1990] (Ont. C.A.) 72 OR (2d) 417
NCC v TB & PB (2014) EWCOP 14
PC and NC v City of York Council [2013] EWCA Civ 478
Re BKR [2015] SGCA 26
Re F (Mental Patient: Sterilisation) [1990] 2 AC
Re GC and Anor [2008] EWHC 3402 (Fam)
Re L (Medical Treatment: Gillick Competency) [1999] 2 FCR 524
Re L (Vulnerable Adults: Court's Jurisdiction) [2014] Fam 1
Re MB [1997] EWCA Civ 3093

Re MP; LBH v GP [2009] Claim No: FD08P01058
Re SA [2005] EWHC 2942 (Fam)
Re SB (A Patient; Capacity to Consent to Termination) [2013] EWHC 1417 (COP)
Rochdale Metropolitan Borough Council v KW & Ors (Rev 1) [2014] EWCOP 45
Saulle v Nouvet [2007] EWHC 2902 (QB)
SCC v LM & Ors [2013] EWHC 1137 (COP)
Sheffield City Council v S [2002] EWHC 2278 (Fam)
V v R [2011] EWHC 822
W v M, S, NHS Primary Care Trust [2011] EWHC 2443 (Fam)
Wandsworth Clinical Commissioning Group v IA [2014] EWCOP 990
Westminster City Council v Manuela Sykes [2014] EWHC B9 (COP)
Wye Valley NHS Trust v Mr B [2015] EWCOP 60
XCC v AA & Anor [2012] EWHC 2183 (COP)

1

Problems with Mental Capacity

Individuals make decisions all the time, ranging from the mundane, such as which cereal we choose to have in the morning, to the profound, such as whom to marry or which religion to follow. The importance of choice in our lives makes the value of autonomy a core pillar of liberal society. Despite its alleged universal importance, the right to make decisions about one's life has been extended to individuals with mental impairments only very recently. With this shift comes a challenge to understand what it means to exercise autonomy in the context of impairment and disability.

Medico-juridical practice and bioethicists commonly posit that individuals with cognitive and psychosocial impairments can exercise their autonomy when they demonstrate mental capacity. At its core, the concept of mental capacity captures the simple intuition that we need to display a level of decision-making competence in order for our choices to be respected; it is a technical concept that assesses the following: can individuals understand and reason about the various options available to them? Can they understand the consequences of their decisions? Are their reasons internally consistent? Can they draw upon true as opposed to deluded beliefs? In short, capacity is an all-or-nothing concept that determines whether others around them defer to individuals' subjective choices, protecting individuals' autonomy from outside intrusion. In many legal jurisdictions, a finding of incapacity means best interests decisions can be made on behalf of another according to what others believe to be beneficial to their well-being, overriding their subjective choices.

Yet this prevalent concept of mental capacity has become increasingly fraught terrain in bioethical and medico-juridical contexts. Consider the following scenarios:

- A woman with Down syndrome can understand why one would use contraception and the implications and risks of not using it; she does not wish to become pregnant. Yet she still refuses to use contraception

due to the coercive influence of her husband, who wants a baby and threatens to leave her if she starts using birth control.
- An elderly woman with mild dementia grants power of attorney to her son, who manipulates, abuses, and isolates her.
- A woman with cognitive impairments is coerced into marrying in a foreign country due to family influence, even as this does not express her own preferences.
- A young man with learning impairments refuses to move away from his enmeshed relationship with his mother, cultivating within him learned helplessness and unhealthy dependence.

These common scenarios raise fundamental questions about how relationships influence mental capacity and the decision-making process. The role of relationships in mental capacity has not been properly considered in theory or practice. The realm of legal practice currently pulls in opposite directions, recognising, on one hand, the *interpersonal* source of capacity: that the assessment of mental capacity can often turn on the relationships surrounding individuals with impairment. Increasing emphasis on supportive decision-making implies that some relationships foster and sustain capacity whilst others undermine it. Conversely, other prominent legal judgments under the Mental Capacity Act 2005 in England and Wales (MCA) assert the *intrapersonal* source of capacity – namely the causative nexus between mental disorder and the inability to decide. Understanding mental capacity in this way captures common law intuitions about individuals as rights-bearers; moreover, it may have procedural justification from a strict legal perspective in that it helps demarcate the boundary between the MCA and inherent jurisdiction in England and Wales.[1]

Even as practice is moving increasingly towards an intrapersonal direction, such an interpretation of mental capacity is unconvincing at a deeper, theoretical level, particularly when we consider the supportive framework that is simultaneously acknowledged as necessary for individuals with impairments to exercise their autonomy and practical reasoning. Thus far, theoretical conceptualisations have yet to situate mental capacity within relationship. Different approaches have stressed how capacity should be situated within discussions of cognitive procedural processes,[2]

[1] See Appendix 1 for a brief overview of the MCA and inherent jurisdiction in England and Wales.

[2] Allen E. Buchanan and Dan W. Brock, *Deciding for Others: The Ethics of Surrogate Decision Making* (Cambridge: Cambridge University Press, 1990).

authenticity,[3] diachronic values,[4] or emotional aptitude.[5] But absent in these discussions is a sustained examination of how *relationships and intersubjective dialogue* bear on capacity – a problem medico-juridical practice increasingly faces but has yet to generate corresponding theoretical reflection in the bioethical and philosophical literature.[6] In sum, the concept of mental capacity needs to move away from its formalistic, individualistic inflection in bioethical literature and medico-juridical practice.

[3] Jacinta Tan et al., 'Competence to Make Treatment Decision in Anorexia Nervosa: Thinking Processes and Values', *Philosophy, Psychiatry, and Psychology* 13:4 (2006): 267–82; Jacinta Tan et al., 'Competence to Refuse Treatment in Anorexia Nervosa', *International Journal of Law and Psychiatry* 26 (2003): 697–707; also see a distinction between 'abilities-based' as opposed to 'pathology-based' approach, made by Gerben Meynen and Guy Widdershoven, 'Competence in Health Care: An Abilities-based versus a Pathology-based Approach', *Clinical Ethics* 7 (2012): 39–44.

[4] Jillian Craigie, 'Competence, Practical Rationality and What a Patient Values', *Bioethics* 25:6 (2011): 326–33.

[5] Louis C. Charland, 'Is Mr. Spock Mentally Competent? Competence to Consent and Emotion', *Philosophy, Psychiatry, and Psychology* 5:1 (1998): 67–81.

[6] There has been some reflection on the impact of relationships in parallel, slightly unrelated debates about voluntariness of informed consent. The standard bioethical account of valid informed consent to participate in research has tended to separate out what we might call 'capacity-related' issues from 'voluntariness-related issues' (see Robert M. Nelson, et al., 'The Concept of Voluntary Consent', pp. 6–16, and Paul Appelbaum, 'Can a Theory of Voluntariness Be A Priori and Value-Free?', pp. 17–18, both in *American Journal of Bioethics* 11:8 (2011)). Capacity-related issues would include the intrapersonal factors that are causally implicated in an individual's ability to consent, whilst voluntariness-related issues include external factors that can affect one's decision-making (i.e. relational coercion and manipulation). Ultimately, this separation is premised on the assumption that decision-making capacities (i.e. what it means to make autonomous decisions in the world) do not necessarily overlap with mental capacity. However, I find this demarcation both puzzling and unconvincing. From a legal perspective, informed consent cases are separate from mental capacity: the former revolve around issues of medical negligence of risks and the provision of proper information about risks, whereas it seems strange to speak of 'informed consent' to reside somewhere of one's choosing, for example. Moreover, mental capacity combines what these debates separate – namely, how impairment and, in some cases, one's relationships can impinge on an individual's ability to make decisions. It might be that someone like Appelbaum would argue that this rests on a mistake that conflates capacity and voluntariness criteria. But from a theoretical perspective, it seems to me an arbitrary move to separate mental capacity criteria from voluntariness criteria, unless one is committed to a highly reductive explanatory framework, which, in my mind, neglects the close interconnection between the two. Mental capacity involves an interrelated network of other concepts, such as autonomy and rationality, especially when we consider the typical criteria for capacity involving pillars such as 'use and weigh' or 'understand and appreciate'. The argument of this book makes clear that I am committed to a more holistic approach to mental capacity, which will indeed encompass the same concerns that are artificially relegated to voluntariness-related criteria.

Motivating this book are two concerns: first, assumptions in medico-juridical practice and bioethical theory cause capacity to be seen as an objective, either-or concept, primarily rooted in internal biological causes. Whether such an account leads to ethically justifiable outcomes in terms of state intervention is questionable: individuals who might be found to lack capacity on these criteria may in fact be able to make their own decisions given a supportive relational environment. Equally, those who choose to remain within abusive, disabling relational contexts are often found to have capacity, causing public institutions to neglect a duty to intervene in such contexts. The second, deeper motivation revolves around concerns about how a shared world with individuals with impairments can be established so that their decisions and actions can be better situated, interpreted, and understood. Those with impairments and disorders are often viewed by society as 'other', in that their behaviour, their choices, or their way of interacting with their environment seem incomprehensible, beyond the bounds of what is deemed 'normal', whatever that term means. And philosophers aren't immune to the impulse to sideline those with impairments. Normative conceptions of rights, autonomy, and reasoning frequently utilise impairment as *contrast* cases, as illustrating *exceptions* to the rule, for how the normative ideal of such concepts looks like for individuals *without* impairments. I want to push back against this sidelining move in rethinking the concept of mental capacity, so as to provide a shared account of its underlying values. The normative conditions of such values will be more broadly construed so as to include individuals with impairments and their unique, embodied interactions with their environment.

This book argues that mental capacity must be conceived of as a relational concept that can be enhanced through intersubjective dialogue. I assume in the first instance that biological causes to impairment can affect decisional capacity – indeed, as the law does according to recent interpretations. However, capacity assessments must also recognise the *relational* constituents of decisional capacity – how these can interact with and worsen biological factors affecting capacity. Most accept that physical accommodations are often necessary to promote the inclusion and autonomy of individuals with impairments, yet relational aspects – that is, the presence or absence of enabling, inclusive narratives – can equally affect one's decision-making abilities and practical agency. Relationships are central to the ability to cope with one's embodied reality; they can enable one's ability to make decisions about one's life. Yet through environmental conditions of abuse, manipulation, or coercion they can also suppress and neglect the necessary skills one needs for daily coping or making authentic choices – much more so if one's impairments necessitate reliance on others. For this reason, I set aside discussion of the biological causes of mental impairments and simply assume its contributing – but not determinative – influence throughout this book. My purpose is to refocus the concept of mental capacity so that, from one end, it considers the reality of certain unique vulnerabilities that emerge out of embodiment and impairment, whilst, from the other end, it galvanises critical reflection on the obligations, duties, and competencies required *on behalf of others* to enable and promote the autonomy of individuals with impairments.

The argument of the book is as follows:

1. An individual's environment, particularly one's surrounding relationships, affects one's ability to make decisions. This is clear when we examine more carefully the theoretical concepts grounding mental capacity, such as rights, autonomy, and rationality. Philosophical accounts are increasingly challenging the individualistic, subjective temper of these concepts. Once we acknowledge that environmental, relational factors do bear on one's decisional capacity, we need further clarification as to what normative criteria should discriminate between those contexts that support and sustain mental capacity from those that undermine it. Without these criteria, capacity assessments that take into consideration individuals' relational context appear arbitrary; judgments as to whether individuals can make decisions about their care, treatment, living arrangements, or the choice to live or die would simply be a matter of whether capacity assessors subjectively approve or disapprove of an individual's relationships. The book argues that supportive environmental, relational features will cultivate 'autonomy competencies' within individuals with impairments, namely, a range of socially acquired perceptual, psychological, emotional, and cognitive skills necessary to engage with the world and make choices in accordance with one's values. Conversely, the absence of supportive relationships and environments, or the presence of abuse, manipulation, and coercion, can fundamentally disable individuals' decisional abilities.
2. The importance of relationships in supportive decision-making entails a range of interpersonal, enforceable duties and practices. This in turn blurs the current legal boundaries separating mental capacity from best interests, as becomes clear when we consider the justifiability of third-party interventions in disabling but freely

chosen relationships. I argue that in situations where the autonomy competencies of individuals are neglected, third-party interventions in disabling relationships can be warranted. Third-party duties to assist and intervene in abusive relationships involving individuals with impairments will be founded on how their decisional capacity and potential for developing autonomy competencies have become compromised within this context, so that individuals can be placed within an environment where their competencies can be encouraged. However, these interventions must be justified and carried out within certain ethical constraints – that is, the same enabling practices that apply to an individual's immediate relationships.

3. I argue that capacity assessments themselves are intersubjectively situated and that the very manner in which these assessments are carried out can have a profound effect on the individual whose capacity is under scrutiny. Capacity adjudications are informed by their particular medico-juridical environment, by their own traditions, preconceptions and therefore are not value-neutral despite their air of objectivity. These assessments themselves are forms of interventions that become part of an individual's context; consequently, they have the potential to enable or disable individuals' decisional autonomy, just as their immediate surrounding relationships. Capacity adjudicators must therefore deploy certain interpretive skills to facilitate understanding rather than misunderstanding of the individual being assessed. These interpretive skills hinge on the exercise of critical reflexivity within the medico-juridical context, where background values and presuppositions in judgments are explicit and open to scrutiny, *even if* the outcome of the capacity adjudication overlaps with what we think is morally defensible.

In sum, the task of the book is twofold: first, to argue for a relational concept of mental capacity; second, to elucidate the ethical characteristics, obligations, and duties incumbent on the surrounding relationships *as well as* capacity assessors in order to contribute to the enablement of individuals with impairments. These claims could help the law develop toward a more relational interpretation of the mental capacity so that medico-juridical assessors and judges can apply current legal concepts in a manner that will enable individuals with impairments to exercise their autonomy.[7] Moreover, the theoretical analysis of the book provides a

[7] It also could have a potentially broader scope, applying to relationships involving not only those with borderline capacity under the law but also victims of abuse who pass the legal

different approach to the mainstream philosophical understanding of key normative concepts, such as autonomy and rationality, so that individuals with impairment are not utilised to illustrate *exceptions* or *outliers*, but rather their unique potential capacities and abilities are better accommodated.

Theories and Methodology

This book adopts an interdisciplinary approach, bringing philosophical reflection to bear on existing legal practice, thus forming the interpretive lens through which I examine the law of capacity. This aspect of my methodology may be controversial. Obviously the law has its own disciplinary conventions and constraints. From one perspective, the book's analysis of case law tries to be sensitive to these conventions but will likely prove unsatisfactory to the legal scholar. From another perspective, however, drawing upon external normative sources is intrinsic to the process of legal interpretation. I want to examine this latter view in further depth here because it shows how the analysis of case law offered in this book might sidestep objections from a purely legal standpoint. More importantly, we might also see how certain conflicting legal judgments about mental capacity, including those lower down, or at the margins of, the legal hierarchy, may be indicative of conceptual disagreement at a deeper, more normative level.[8]

According to Ronald Dworkin, legal practice and analysis requires an 'interpretive attitude' in which personal and institutional morality interacts.[9] Legal interpretation is constructive in so far as it advances the purposes of the law, yet is also subject to both *internal* and *external* normative constraints. Internal standards of law include the importance of history, precedent, and available interpretations, whilst external

threshold of mental capacity, in the context of promoting their potentialities and abilities in mundane day-to-day settings. I do not discuss this implication in the book, as this would detract from my primary focus here.

[8] I mean 'normative' in the sense that there are certain guidelines that recommend what ought to be the case. In this sense, the normative constraints of the law are not necessarily coextensive with normative *ethical* constraints. Normativity should not be conflated with morality – the latter concept of course will contain normativity, but the opposite is not true, or would require some philosophical manoeuvring to establish.

[9] I set aside Dworkin's arguments about 'law as integrity', which presupposes that judges make interpretive decisions based on the assumption that these laws are agreed upon within the community, in accordance with fairness and justice. I do not believe one has to be committed to this more robust conception of the law to nonetheless draw upon the overall structure of Dworkin's account of the judicial interpretive endeavour.

constraints refer to socio-cultural norms and moral and political ideas. Hard cases or conflicting judgments at the level of law are often indicative of disagreement at a conceptual level, where further examination of the underlying concepts informing them is warranted. This analysis is not just descriptive but involves normative assessments based on (i) *fit* with previous cases, the legal record, and particular legal practices, as well as (ii) *justification* from the standpoint of political morality.[10]

Given that earlier decisions have 'gravitational force',[11] the first criterion of fit may well recommend legal interpretations that are 'not too novel'.[12] As Dworkin states, '[a]t the first level agreement collects around discrete ideas that are uncontroversially employed in all interpretations'.[13] However, the second criterion could 'set these reasons against . . . *more substantive political convictions* about the relative value of two interpretations'.[14] One interpretation could emerge as superior from this standpoint if it 'makes the legal record better overall . . . *even at the cost of the more procedural values.*'[15] In short, the process of normative justification leads to questions of how moral and political theory supports the body of laws.

Dworkin's multileveled methodology is adopted in this book to explore the *normative* concept of mental capacity as it applies to medico-juridical practice. Legal analysis comprises the first level, taking into consideration the evolution of mental capacity case law through specific judicial decisions, keeping in mind the priority of some cases and interpretations over others. The next two levels involve *extra-legal* steps: disagreements in hard cases require judgement to probe additional layers relevant to the case. This doesn't mean that judges are *creating* new law or that they are adding these layers consciously. However, the use of judgement involves engaging in different levels of conceptual abstraction. According to Dworkin, 'the controversy latent in this abstraction is identified and taken up' in subsequent levels, and 'exposing this structure may help to sharpen argument and will in any case improve the community's understanding of its intellectual environment'.[16]

The level of analysis will determine our view as to whether medico-juridical debates about mental capacity are resolved. From a strict

[10] Ronald Dworkin, *Law's Empire* (Oxford: Hart, 1998), Dworkin, 'Natural Law Revisited', *University of Florida Law Review* 34:2 (1981–2): 165–88.

[11] Ronald Dworkin, *Taking Rights Seriously* (London: Duckworth, 1996), p. 111.

[12] Dworkin, *Law's Empire*, p. 248. [13] Ibid., p. 71.

[14] Ibid., p. 248, emphasis added. [15] Ibid., emphasis added. [16] Ibid., p. 71.

legal perspective, certain issues of mental capacity may revolve around consensus *internal* to the *practice* of law. We might think that developing case law and the clarification of procedural boundaries between capacity law and inherent jurisdiction together point to consensus that mental capacity is an intrapersonal concept. However, Dworkin's legal interpretivism suggests that these internal resolutions *by themselves* do not decide the *conceptual and ethical issues* underlying hard cases about mental capacity. Analysis at the first level does not exhaust what has to be said about concepts operating within the law; it may even indicate contested or incoherent assumptions. For this, a second level of analysis is needed, drawing on phenomenology, for example, to examine how legal interpretations cohere with or depart from the lived experience of impairment and presenting to us a more nuanced understanding of how important concepts, such as autonomy, reasoning, impairment, and disablement, manifest themselves in lived practice. The final level is normative, to provide a phenomenologically sensitive yet justifiable account of the values and concepts at stake in the medico-juridical practice of mental capacity.

I have not devoted separate sections of the book to each of procedural issues, phenomenological analysis, or normative argument but incorporate these different levels of analysis in a more holistic fashion. In the first instance, my argument draws on an eclectic range of philosophical sources, and my reasons for favouring some theories over others will become clear in each chapter. This philosophical eclecticism is important, as I would fail to do justice to the different ethical perspectives that warrant consideration if I endeavoured to apply a grand unifying theory. First-personal, second-personal, and third-personal dimensions of mental capacity need to be taken into account. The first-personal dimension involves the subjective experience of capacity. Phenomenological analysis will therefore be crucial to account for the first-hand experience of embodiment and impairment to offset the priority accorded to cognitive reflection in our conception of autonomy, which often excludes individuals with impairments. The second-personal dimension denotes the relational, intersubjective constituents of capacity – the type of interpersonal engagement, duties, and obligations that are owed to us by others. In this respect, feminist theory helps illuminate the interpersonal constituents in the development of autonomy, as well as challenges misguided individualistic, liberal assumptions implicit in our notion of legal rights. Hermeneutics moreover allows us to articulate more fully the substantive intersubjective practices that underlie enabling relationships and capacity assessments.

Finally, the third-personal dimension refers to the perspective of others who stand outside our immediate relationships. For this, a more deontological, Kantian approach captures the unique normative, third-party standpoint of capacity adjudications, whilst more substantive theories of rationality highlight the socially embedded nature of these judgements themselves.

As a whole, this multileveled, multi-perspectival methodology draws attention to instances where a more relational concept of mental capacity is or ought to be assumed in practice; it likewise pinpoints where our theoretical understanding and analysis of contested background values and concepts could improve. Some degree of 'reading between the lines' inevitably occurs in my analysis of the law and case reports, but this is necessary if we are to ensure that mental capacity as a medico-juridical concept is properly responsive to the realities and challenges of exercising autonomy within the context of impairment.

Some might question the validity of this methodological approach on grounds that legal analysis eschews the phenomenological and normative ethical levels of analysis. If we are to draw any normative conclusions about cases, these should surround issues about procedural mechanisms and judicial reasoning within the confines of the legal discipline and its traditions. The *legal* significance of specific cases is not coextensive with their *normative* significance. The naturalistic fallacy is committed when we conflate the two, thus confusing descriptive and normative endeavours. If true, the normative ambitions of this book would have to be scaled back dramatically.

Awareness of the constraints of legal analysis is important, particularly if my argument is to convince medico-juridical practitioners and legal scholars. Yet I will not confine myself to the first level of analysis for three important reasons. First, a legal-procedural analysis would overdetermine the priority of an intrapersonal, individualistic reading of mental capacity, particularly as legal precedent more recently has begun to sideline issues having to do with the impact of relationships on individuals' ability to decide, assigning them to legal mechanisms *outside* mental capacity law, that is, the court's inherent jurisdiction in England and Wales. Focusing on legal procedure – and how jurists are interpreting these procedural boundaries – unduly restricts the questions we choose to ask and the answers we give about mental capacity.

Second, the objection artificially restricts the scope of legal analysis. To be clear, I do not claim that normative insights can be *derived directly* from what is said in legal judgments, nor that *the cases themselves* bear the full weight of my normative analysis. Yet legal judgments do not occur within a vacuum separated from moral insights and political concerns; they change alongside developments in society and our understanding of individuals' lived experience. Attempts to hive off 'extra-legal' concerns from the practice of judicial deliberation have already been exposed as disingenuous in legal scholarship. For example, feminist jurists convincingly show how legal analysis by itself is insufficient to critically assess assumptions within existing legal mechanisms that entrench questionable practices of gender inequality, patriarchy, and violence against women.[17]

Finally, this objection assumes that descriptive and normative endeavours are one and the same in my methodology, where the content of certain practices is presumed coextensive with the content of a particular rule. My argument would indeed fall foul of the naturalistic fallacy if this were true. My methodology assumes instead that particular practices help *justify* a normative judgement or rule. There is a degree of separation between the *existence* and *value* of a practice, meaning their respective content could diverge.[18] That said, practices cultivate expectations about our duties and how these are to be met, and in this way help legitimise normative claims. In turn, these expectations and normative claims shape our sense of what practices are required, of where their purpose lies.[19] The law has an analogous overlapping interpretive-normative structure: legal practice identifies commonly held rules and standards, but greater levels of abstraction are required when considering its justification. These normative insights are then applied to legal practice to improve its rules and standards. Normative judgement invariably takes interpretations of practice as its starting point. So pure legal analysis might focus on gleaning a plausible interpretation of mental capacity based on where specific cases sit in the legal hierarchy or the strength of the judicial reasoning. But understanding the relational constituents of mental capacity can emerge only if the separation between descriptive and normative content is maintained whilst acknowledging the evolving interaction between political, ethical norms and medico-juridical practice.

17 See Jennifer Nedelsky, *Law's Relations: A Relational Theory of Self, Autonomy, and Law* (New York: Oxford University Press, 2013); Nicola Lacey, *Unspeakable Subjects; Feminist Essays in Legal and Social Theory* (Oxford: Hart, 1998).

18 Dworkin, *Law's Empire*, p. 57.

19 In Dworkin's schema, he calls this the 'postinterpretive or reforming stage', see ibid., p. 68.

A Note on Terminology, Philosophical Concepts, and Legal Jurisdiction

Too often the language surrounding those with impairments has been discriminatory and demeaning, with the suggestion that they are undeserving of respect, dignity, and resources. Disability rights activists and social models of disability rightly highlight how language can contribute to these pernicious views. This book will not subscribe to all the premises of the social model, but their distinction between 'impairment' and 'disability' illuminates how the very term 'disability' is already a determinative judgement, conjuring up images of individuals who somehow fall short. The language that constitutes our surrounding narratives has a profound impact on how we see ourselves, how we position ourselves vis-à-vis others. For this reason, throughout the book I try to refer to those subject to capacity adjudications as 'individuals with impairment' rather than 'disabled' individuals. This isn't to say that all individuals with impairments are the same or that they share the same degree or kind of impairment. But I adopt the terminology in hopes that it is elastic enough to encompass a range of impairments and levels of individual functioning.

The argument in this book draws on a number of important philosophical concepts, such as autonomy, rationality, rights, and beneficence. In the first instance, I recognise that numerous interrelated philosophical debates are likely relevant to my discussion, but out of necessity, I have restricted the scope of these concepts. Take, for example, the concept of autonomy: it is often the case that philosophical examinations of autonomy are bound up with debates about moral responsibility – the extent to which one can be held culpable for one's actions and intentions. The case for respecting autonomy typically hinges on competence to be held morally responsible, and how we conceptualise the threshold of moral responsibility has profound consequences for whether the choices of individuals with impairments are given due respect.[20] Moral responsibility further implicates issues about the type of reasoning and responsiveness to reasons one requires. Likewise, autonomy relates to disagreements about compatibilism – the extent to which one is free to do otherwise and whether freedom of the will is in some way compatible with determinism. I do not address these important philosophical

[20] See Camillia Kong, 'The Space between Second-Personal Respect and Rational Care in Theory and Mental Health Law', *Law and Philosophy* 34 (2015): 433–67

questions in the book, although I recognise they may be connected to my overall argument in some fashion. I sometimes try to address concerns my argument may provoke in these related debates, but out of necessity I am selective about the theoretical issues I engage in order to navigate such messy philosophical terrain.

The case studies of the book focus primarily on the case law in England and Wales, mainly the Mental Capacity Act and the use of inherent jurisdiction. The developing case law is at the forefront of interpretive issues surrounding mental capacity. The MCA symbolises the formal (if imperfect) extension of the autonomy ideal to individuals with impairments.[21] According to Lord Falconer, the MCA aims to 'empower people to make decisions for themselves wherever possible, and protect people who lack capacity by providing a flexible framework that places individuals at the very heart of the decision-making process.'[22]

The issues that crop up within the MCA are a microcosm of broader concerns that are not jurisdiction-specific. Cases about relationships and their impact on the autonomy of individuals with impairments manifest themselves time and time again. As the autonomy ideal increasingly encompasses all individuals, regardless of impairments, the distinction between relationships that can help, assist, and enable, as opposed to those that neglect, abuse, and disable, will become all the more important. These issues replicate themselves in other legal spheres, such as the oft-debated question about the specific supported decision-making mechanisms contained within the United Nations Convention for the Rights of Persons with Disabilities (CRPD). Future mental capacity (or indeed, legal capacity) legislation may eventually resolve questions of how the law ought to recognise the role of relationships in developing and promoting capacity. But the manner in which the courts in England and Wales have tried to arbitrate this question will be an instructive point of departure for more theoretical discussions, especially as the MCA has been used as a template elsewhere.[23] That said, I am also aware that the MCA is far from perfect and recognise the need to engage in key debates emerging out of the CRPD in particular.

[21] The MCA is controversial from the Conventions on the Rights of Persons with Disabilities perspective particularly for its diagnostic threshold and retention of the best interests, substituted decision-making framework.

[22] Department of Constitutional Affairs, *Mental Capacity Act 2005 Code of Practice* (London: TSO, 2007), foreword.

[23] I.e. Singapore Mental Capacity Act.

Structure of the Book

Each chapter of the book examines a different facet of a relational concept of mental capacity. Chapter 2 builds the case that legal rights already structure relationships in particular ways. I start out by providing a brief explanation of the legal context and bioethical heritage of the concept of mental capacity. Whilst a human rights–based perspective increasingly influences approaches to disability, this has not led to a unified approach to mental capacity law nor to bioethical theoretical discussions of capacity. In short, there are *internal* and *external* challenges to the concept: on one hand, applications of the MCA reflect contradictory views as to whether – and to what degree – decisional capacity is determined by intrapersonal or relational factors, resulting in discrepant judicial interpretations. On the other hand, the advent of the 'will and preferences' paradigm through the CRPD's claim of universal capacity presents an external challenge to the concept of mental capacity, whilst advocating the importance of supportive mechanisms for individuals with impairments. Chapter 2 focuses on the 'will and preferences' paradigm in depth and argues that its critique of mental capacity is problematic because of its adherence to two questionable assumptions in their interpretation of rights: (i) the private/public distinction and (ii) liberal individualism. The chapter then draws on feminist legal scholarship examining the law's complicity to violence against women to defend a relational approach to legal rights. This shifts the analytical focus toward how rights are designed to *structure relationships* and challenges the peremptory authority of individual persons typically assumed in liberal rights. Moreover, it suggests how a relational analysis of rights provides a crucial prism through which the law can recognise how relationships enable or disable individuals' mental capacity.

Chapter 3 argues that a relational concept of mental capacity will demand reconceiving our underlying account of autonomy. Mental capacity legislation is grounded in normative concerns about respect for the autonomy of individuals with impairments. However, a certain degree of imprecision and inarticulacy tends to characterise current legal appeals to respect for autonomy. This chapter argues that autonomy encompasses crucial perceptual and reflective skills: the phenomenology of absorbed coping draws attention to how impairment can affect one's perceptual engagement with the world, yet unavoidably encourages social and physical engagement. This reiterates how autonomy competencies, comprising socially acquired skills of self-understanding and self-reflection, rely on a network of interpersonal support. In the context of capacity, this chapter provides *interpersonal* and *intrapersonal* indicators of autonomy-enabling relationships, focusing on whether the sociocultural context permits and encourages first-personal revisions and whether the individual has the confidence, skills, and self-esteem necessary to initiate and endorse these revisions. Moreover, concerns about unwarranted perfectionism are mitigated because the account endorsed in this chapter conceives of autonomy more as a matter of degree.

Chapter 4 probes the account of rationality that ought to underlie a relationally situated account of mental capacity. It presents an internal and external critique of the procedural model of rationality that is commonly presupposed in the functional test of capacity, arguing that this model is not fit for purpose in light of the theoretical assumptions of relational autonomy. This model's explanatory and normative minimalism is defective: the procedural model's theoretical assumptions surrounding rational justification are shown to be untenable without presupposing a substantive, social account of rationality. Moreover, its rational norms fail to provide the evaluative tools needed to recognise rationally consistent choices that result from coercive relational dynamics. For such a task in capacity adjudications, an alternative mode of reasoning is needed, one that allows and facilitates the explicit articulation and evaluation of one's substantive reasons, of how these reasons draw on or reinterpret certain intersubjectively shared values. I follow Charles Taylor in calling this the 'articulatory-disclosive' mode of reasoning. Unlike the procedural model, the articulatory-disclosive mode of reasoning can (i) express the medico-juridical value of promoting an individual's autonomy competencies and (ii) engage with the intrinsically substantive question of when decisional support melds into coercive duress.

Chapter 5 explores how a relational concept of mental capacity generates interpersonal duties of support and intervention. The social model of disability is commonly thought to justify the duties of interpersonal support that are expressed in the CRPD's supportive decision-making mechanisms. However, the theoretical resources of the social model are ill equipped for the normative justification of positive duties that respond to undesirable harms intrinsic to one's impairment or the ways in which certain impairments make one more vulnerable to mistreatment in relationships. Similarly, what is morally required in cases where relational abuse and neglect contributes to an individual's disablement is ambiguous but suggests that intervention would be strictly impermissible. The chapter adopts a Kantian approach to reflect on the particular normative perspective of capacity assessors who may be charged with carrying out positive duties. Kant's account of practical reasoning contains important normative insights for this purpose, given that positive duties of assistance emerge out of a gap between abilities and aspiration. Not only does this approach lessen the conceptual divide between impaired and non-impaired individuals, it also provides crucial moral justification of the interpersonal duties constitutive of supportive decision-making, such as developing and promoting autonomy competencies in others. It also responds to questions of when outside interventions in disabling relationships are ethically warranted.

Chapter 6 articulates the dialogical practices of those relationships, capacity assessments, and third-party interventions that enable individuals' mental capacity. Relational conditions of asymmetrical responsibility challenge the dichotomy between 'respect' and 'care' in traditional moral theory. The concept of hermeneutic competence shifts the ethical attention away from the skills of individuals with impairment towards the competencies of those with whom they interact. Both internal practices and external interpretations of capacity need to display three features of hermeneutic competence – (i) phenomenological awareness, (ii) dialogical understanding, and (iii) self-nurturing recognitional mechanisms – in order to promote the autonomy of individuals with impairments. This discussion also explores how medico-juridical assessments of capacity are in a unique position to initiate narrative repair where appropriate (which makes space for and supports narratives of resistance against those that fundamentally damage individuals' personal identity). Ultimately, the same dialogical practices that promote capacity in the immediate relational context should likewise describe the appropriate ethical stance of the medico-juridical professional assessing capacity.

Finally, Chapter 7 evaluates the practical and theoretical implications of a relational concept of mental capacity. It summarises the characteristics of relationships that enhance an individual's autonomy competencies and practical reasoning. This subsequently helps provide crucial normative criteria through which practitioners in the medico-juridical context can discriminate between the kinds of communities that do and do not support decisional capacity and further elucidates the ethical foundation for supportive decision-making mechanisms demanded under the CRPD. The conclusion, then, is that the concept of mental capacity – far from being a technical concept merely requiring application – is an intrinsically *interpretive, dynamic, and dialogical concept* that draws on a range of implicit values and goods; chief amongst them is the importance of socially acquired autonomy competencies needed for temporally extended agency. Even more significantly, mental capacity understood as such shifts the burden of practical understanding away from the individual in question, towards the ethical-dialogical stance of those around the individual, including the assessor.

Jonathan Herring argues, 'it is simply impossible to make an assessment of any individual's interests outside the context of their relationships'.[24] The goal of this work is precisely to advance this case for mental capacity within relationship, to capture the characteristics of relational support that enables rather than disable individuals with impairments, and to reorientate our ethical attention to the various competencies and practices of the communities that surround them.

[24] Jonathan Herring, *Caring and the Law* (Oxford: Hart, 2013), pp. 155–6.

2

Mental Capacity, Legal Capacity, and Relational Rights

I. The Shifting Landscape

The value of autonomy in the medico-juridical sphere is a well-established cornerstone for consent to clinical treatment and research. Importantly, the threshold of mental capacity 'function[s] as a gatekeeping device for informed consent' and officially demarcates spheres of autonomy and paternalistic care in many legal jurisdictions. Not all definitions of mental capacity are created equal, however. The Law Commission rejected status-based or outcome-based tests of capacity in early debates during the Mental Capacity Act's (MCA's) inception. Both tests of mental capacity have fallen out of favour in medico-juridical contexts for obvious reasons. Status-based tests were deemed discriminatory due to their reliance on a person's diagnosis to make blanket judgements about capacity, making it 'quite out of tune with the policy aim of enabling and encouraging people to take for themselves any decision which they have capacity to take'.[1] Meanwhile outcome-based tests tended to ground capacity on whether an assessor agreed with the decision or its underlying values, thus 'penalis[ing] individuality and demand[ing] conformity at the expense of personal autonomy'.[2] Many jurisdictions consequently adopt a *functional test* that allegedly tracks the structure of reasoning or process of thought required to make time- and context-specific decisions.[3] In theory, this simultaneously mitigates issues that plague other tests of capacity, thus protecting the autonomy of individuals with impairments whilst installing substituted decision-making mechanisms that safeguard the welfare of those lacking capacity. Proponents argue that the MCA

[1] Law Commission, *Mental Incapacity; Item 9 of the Fourth Programme of Law Reform: Mentally Incapacitated Adults* (Law Com 231, 1995), 3.3.

[2] Ibid., 3.4.

[3] I.e. MCA in England and Wales, the MCA in Singapore, much of Canadian law (i.e. Health Care Consent Act, 1996, Substitute Decisions Act, 1992, The Adult Guardianship and Co-decision-making Act, SS. 2000).

represents 'a new approach to capacity' grounded on 'the philosophy... that those who suffer from disability shall be assisted to live normal lives and to make choices about those lives to the greatest extent possible.'[4]

But functional tests of mental capacity are not immune from criticism either. From an internal perspective, its underlying conception of autonomy and rationality can be problematised, as we will see in the subsequent chapters. From an external perspective, mental capacity as a legal and ethical concept has itself become questioned. The United Nations Convention for the Rights of Persons with Disabilities (CRPD) represents a major shift in this respect, arguing that the rights of individuals with impairments demand the further protection of individuals' right of autonomy, opposing the presumption that such individuals can be subject to paternalistic care or substituted decision-making. Instead, states and institutions are obliged to provide supportive structures and mechanisms to ensure individuals with impairments can make their own decisions about their care and treatment.

What I want to do in this chapter is critically examine in more detail the possible external challenge posed by the CRPD, particularly the influential 'will and preferences' interpretation of Article 12.[5] Closer analysis of the 'will and preferences' paradigm reveals shortcomings with this critique of mental capacity, particularly in its response to issues of relational abuse and coercion. This chapter makes two arguments: first, the interpretation of rights implicit in the 'will and preferences' paradigm rest on questionable liberal assumptions about the (i) private/public divide (which demarcates spheres where the state can or cannot intervene) and (ii) the individualistic nature of rights (which posits that rights prioritise and protect the subjective claimant).[6] Second,

[4] *Saulle v Nouvet* [2007] EWHC 2902 (QB), para.23.

[5] It is important to note that the 'will and preferences' paradigm, though influential, does not necessarily reflect the convention obligations of the CRPD.

[6] When I use the term 'liberal' here and elsewhere in the book, I am referring to the *political philosophy of liberalism* and its body of commitments as opposed to its practical enactment in the political spectrum, which can vary depending on country and context. The liberal tradition contains a number of commitments, such as the protection and value of individual liberty and civil rights against incursions of the state and the importance of democratic decision-making and transparent public institutions. The liberal school is broad and contains conflicting theories about redistribution, justice, fairness, the role of state welfare and culture, and conceptions of the self. Thus, broadly speaking, one political doctrine that endorses limited state and private property, and another that favours provisions of state welfare and higher taxation, both may fall under the broad rubric of philosophical 'liberalism', even as our everyday usage of the term 'liberal' denotes where one lies within the political spectrum.

internal and external critiques converge in drawing attention to ways in which mental capacity as a concept demands a relational account of rights. Drawing upon work by feminist legal scholars, I suggest that a relational analysis of rights attends to ways that the law and rights structure relationships, providing us with crucial analytical tools to critically examine the types of relationships and underlying values that are normatively desirable and focus on how these can be realisable in law.

The structure of the chapter is as follows. Section II outlines briefly the internal problems with the individualistic assumptions of the functional test of mental capacity, to be discussed in subsequent chapters. Section III discusses the CRPD's potential critique of mental capacity, focusing on the 'will and preferences' interpretation of legal capacity and Article 12. However, this interpretation's implicit adherence to assumptions about the private/public divide and individualistic rights are problematic in abusive contexts involving individuals with impairments, as I discuss in Sections IV and V. As a result, I remain sceptical of whether the 'will and preferences' paradigm is a viable substitute for the concept of mental capacity. To mitigate both internal and external critiques, I explore a relational approach to rights as a crucial legal prism to interpret mental capacity.

II. Mental Capacity and the Internal Challenge

The MCA test of mental capacity has two stages: first, individuals must meet the *diagnostic threshold* and evidence a disturbance of the mind (temporary or permanent). Second, a functional, time- and decision-specific test of capacity determines whether an individual has capacity to make decisions about his or her care and treatment. The functional focus means that a particular diagnosis cannot determine the capacity of individuals; its time- and decision-specific nature means capacity must only be evident 'at the material time' of making a decision, notwithstanding an individual's ability to make long-term plans. Based loosely on competency tests of valid informed consent,[7] the MCA adjudicates capacity according to four criteria. These include the ability to:

[7] As outlined by Paul S. Appelbaum and Thomas Grisso, competency is defined by four categories under the Mac-CAT-T as follows: (i) expressing a choice, (ii) understanding, (iii) appreciation, and (iv) reasoning. See their papers, 'The MacArthur Treatment Competence Study. I: Mental Illness and Competence to Consent to Treatment', *Law and Human Behavior* 19:2 (1995): 105–26; 'The MacArthur Treatment Competence Study. II: Measures of Abilities Related to Competence to Consent to Treatment', *Law and Human Behavior* 19:2 (1995): 127–48; Thomas Grisso et al., 'The MacCat-T: A Clinical Tool to Assess Patients' Capacities to Make Treatment Decisions', *Psychiatric Services* 48:11 (1997): 1415–19.

1. understand information relevant to the decision,
2. retain that information,
3. use and weigh that information, and
4. communicate a decision.

These criteria are presumed to track the basic procedural reasoning skills required to make decisions without violating commitments of liberal neutrality. At least in theory, it remains impartial towards the decision's content, values, or consequences.

Yet the functional criteria of mental capacity are vulnerable from both *internal* and *external* challenges. I will only discuss the internal challenges briefly here, as subsequent chapters discuss these in much more depth. Most capacity adjudications hinge on the third pillar, the ability to 'use and weigh' information. As Peter Bartlett states, '[w]hen the courts have made controversial findings of incapacity in the past, it has often been as a result of applying this rather amorphous standard'.[8] 'Use and weigh' allegedly captures basic reasoning skills, such as consequential and comparative thinking and adherence to rules of logic. Moreover, it has been interpreted as a process occurring within an individual's mind, in relative isolation from others. The MCA states in S.1(3) that '[a] person is not to be treated as unable to make a decision unless all practicable steps to help him to do so have been taken without success', but what these 'practicable steps' entail is unclear. Some critics charge that the domain of law assumes an individualistic conception of capacity that heavily privileges medical expertise, thus departing from standards of clinical practice where capacity assessments tend to be more relational and dialogical: 'Although competence is a matter of a dynamic or dialogue between doctor and patient,' Stefan states, 'legal doctrine sets up this dialogue so that the powerful half of the conversation remains entirely invisible'.[9]

This emerges in two ways: first, overdependence on medical expertise contributes to the minimisation of the voice of individuals whose capacity is being assessed; second, the idea that capacity reflects intrapersonal cognitive skills ignores how the relational context, as well as other non-cognitive skills, can contribute to or detract from these decision-making abilities. As Herring rightly advocates, '[a]t least part of the assessment

[8] Peter Bartlett, *Blackstone's Guide to the Mental Capacity Act 2005*, 2nd ed. (New York: Oxford University Press, 2008), p. 51.

[9] S. Stefan, 'Silencing the different voice: competence, feminist theory and law', *University of Miami Law Review* 47 (1993): 783, quoted in Mary Donnelly, 'Capacity assessment under the Mental Capacity Act 2005: Delivering on the functional approach?' *Legal Studies* 29:3 (2009): 480.

of capacity should be the extent to which the person within their support group of family and/or friends is able to make choices [and] give sufficient weight to the way that others can enable the person lacking capacity to make a decision'.[10] Legal applications of the functional test often make two philosophically contestable assumptions: (i) autonomy is an individualistic rather than relational concept; (ii) capacitous reasoning is an intrapersonal rather than interpersonal act, reflective of the individual's own cognitive processes. Both these assumptions are inattentive to ways in which autonomy and rationality are vulnerable to internal *as well as external* compulsions. Oppressive social relationships, pressures outside the self, and different 'spaces of reasons' can all compromise the full range of capacitous agency, from basic procedural reasoning skills to the ability to make authentic decisions. These problems will not be resolved here but are the focus of Chapters 3 and 4.

II. The CRPD and the External Challenge

External critiques likewise challenge the concept of mental capacity.[11] Most notably, the CRPD contests the contingent link between decision-making abilities and legal agency at the heart of mental capacity regimes. Prominent interpretations of the CRPD make two claims: first, the notion of 'mental capacity' as a whole is a discriminatory concept; second, substituted decision-making regimes (where decisions are made on behalf of those found to be 'incapacitous') violate the human rights of individuals with impairments. The Committee on the Rights of Persons with Disabilities states explicitly that 'perceived or actual deficits in mental capacity (or decision-making skills) must not be used as justification for denying legal capacity'.[12] It is fairly obvious why status- or outcome-based tests of capacity are objectionable – the former tends to discriminate on the basis of a person's diagnosis, whilst the latter deploys third-party value

[10] Herring, *Caring and the Law* pp. 157, 158.

[11] The remainder of the chapter reproduces large sections from my article, 'The Convention for the Rights of Persons with Disabilities and Article 12: Prospective Feminist Lessons against the "Will and Preferences" Paradigm', *Laws* 4 (2015): 709–28. Some modifications here alter the original text.

[12] United Nations Committee on the Rights of Persons with Disabilities, 'General Comment No. 1 – Article 12: Equal Recognition before the Law UN Doc. No. CRPD/C/GC/1, adopted at the 11th Session', (April 2014), para. 13, available online: http://www.ohchr.org/EN/HRBodies/CRPD/Pages/GC.aspx (accessed 6 October 2015). It is important to note that, although influential, statements by the Committee on the Rights of Persons with Disabilities, particularly 'General Comment No. 1', are contested.

judgements about the consequences of individuals' choices. Yet the Committee on the Rights of Persons with Disabilities also understands the CRPD as rejecting the legal and ethical validity of functional tests, stating that they are: '(a) discriminatorily applied to people with disabilities; and (b) . . . presum[e] to be able to accurately assess the inner-workings of the human mind and, when the person does not pass the assessment, it then denies him or her a core human right – the right to equal recognition before the law'.[13]

On these grounds, proponents interpret the CRPD as recommending the replacement of mental capacity with legal capacity.[14] Profound implications in terms of the rights, entitlements, and obligations that are owed to individuals with impairments follow as a result. Legal capacity is grounded on (i) the protection of *formal equality* and (ii) provisions to secure *substantive equality*.

Formal equality refers to the fair treatment of individuals irrespective of contingent features of their personhood, such as ethnicity, gender, and physical or personal attributes. Equal treatment in this respect requires abstracting from an individual's defining characteristics, focusing instead on universal, value-neutral features we all share, such as our common humanity.[15] Formal equality in this sense forms the backbone to the concept of universal legal capacity. Unlike the notion of mental capacity, legal capacity is intrinsic to all individuals, regardless of decision-making abilities, and is a 'universal attribute inherent in all persons by virtue of their humanity'.[16] Article 12 further states that 'persons with disabilities enjoy legal capacity on an equal basis with others in all aspects of life'.[17] Importantly, such formal equality entails that individuals with disabilities are rights-bearers with legal agency that deserves state recognition,[18] thereby

[13] Ibid., para. 15.

[14] See Emma Cave, 'Determining Capacity to Make Medical Treatment Decisions: Problems Implementing the Mental Capacity Act 2005', *Statute Law Review* 36:1 (2015): 99.

[15] I set aside the issue of the 'status' of individuals with severe cognitive impairments. Suffice to say that proponents of the CRPD (and the normative intent of the CRPD) seek to challenge claims that those with significant impairments lack the status of personhood. But for more of this debate, see Eva Feder Kittay, 'At the Margins of Personhood', *Ethics* 116 (2005): 100–31; Eva Feder Kittay, *Love's Labor: Essays on Women, Equality, and Dependency* (New York: Routledge, 1999); the works of Jeff McMahan: *The Ethics of Killing: Problems at the Margins of Life* (New York: Oxford University Press, 2003); 'Cognitive Disability, Misfortune, and Justice', *Philosophy and Public Affairs* 25 (1996): 3–35; 'Radical Cognitive Limitation', in K. Brownlee and A. Cureton, eds., *Disability and Disadvantage* (Oxford: Oxford University Press, 2009), pp. 240–59.

[16] 'General Comment No. 1', para. 8.

[17] CRPD, Art. 12.

[18] See 'General Comment No. 1', para. 5.

signalling 'the change from welfare to rights' where 'the equality idiom [applies to] both same and different persons with disabilities'.[19] This marks an important shift from mental capacity regimes where, although individuals may be entitled to certain basic rights, their legal agency – such as the right to make legally binding decisions in contracts, about marriage, treatment, and care arrangements – tends to be contingent on them meeting a certain threshold of decision-making competence. By contrast, emphasis on the person with impairment as rights-bearer helps guarantee their legal status and agency on par with able-bodied individuals.[20]

But the formal equality embedded within the concept of legal capacity can only be secured through provisions of *substantive equality*. Whereas formal equality centres on commonality and the universal, substantive equality requires a more particularistic, contextualised focus so that existing institutional supports and resources are tailored specifically to the differential needs of individuals and groups. The same treatment of certain individuals by societal and institutional structures can signal a failure to treat them equally in a substantive sense. For example, consider how public spaces have the potential to include and exclude. The use of steps rather than slopes, or the absence of elevators in public transport, illustrate how the same treatment of individuals has differential, unequal consequences: the assumption that we are all the same (i.e. able-bodied) effectively functions as barriers to the equal treatment, access, and inclusion of those with physical impairments. Thus, a crucial part of the CRPD is the protection of substantive equality so that institutions must provide 'reasonable accommodations' to individuals with impairments.[21]

Supportive mechanisms must be available to secure both the formal and substantive equality of such individuals. Decisional support implies a range of macro- and micro-duties, extending from state policies and legislative changes, to interpersonal advocacy and networks of support.[22] Most importantly, Article 12 contains the crucial clause that the 'exercise of legal capacity' must be ensured to 'respect the rights, will and preferences of the person'.[23] This signals the rejection of substituted decision-making mechanisms where other individuals can make best interests decisions on

[19] A. Dhanda, 'Constructing a New Human Rights Lexicon: Convention on the Rights of Persons with Disabilities', *Sur – Revista Internacional de Direitos Humanos* 5 (2008): 45.

[20] Michael Bach and Lana Kerzner, *A New Paradigm for Protecting Autonomy and the Right to Legal Capacity* (Ontario: Law Commission of Ontario, 2010), p. 31.

[21] 'General Comment No. 1', para. 33.

[22] See Piers Gooding, 'Supported Decision-Making: A Rights-Based Disability Concept and Its Implications for Mental Health Law', *Psychiatry, Psychology and Law* 20 (2014): 431–51.

[23] CRPD Article 12 (4).

behalf of another with a finding of mental incapacity. According to the Committee on the Rights of Persons with Disabilities, the principle of 'best interests' is non-compliant with the 'will and preferences paradigm' and violates the right of individuals with impairments to enjoy legal capacity 'on an equal basis with others'.[24] This accordingly 'opens up zones of personal freedom' and 'facilitates uncoerced interactions', which enables individuals with impairments to exercise their right of autonomy and self-determination.[25]

There are strict and flexible interpretations of the 'will and preferences paradigm'. The General Comment of the Committee on the Rights of Persons with Disabilities signals a strict interpretation that rejects any justification for best interests decision-making on behalf of individuals with impairments. Respect for their legal capacity is coextensive with deference to their rights, will and preferences with regards to their choices about health, treatment, and care. In short, the subjective preferences of the individual are prior to any other welfarist considerations or third-party obligations to intervene. Michael Bach and Lana Kerzner, in their report for the Ontario Law Commission, offer a more flexible interpretation of Article 12, proposing three decision-making statuses. *Legally independent status* refers to those individuals who possess 'requisite decision-making abilities' that are tracked by 'understand and appreciate' tests traditionally associated with tests of mental capacity (i.e. the ability to understand, retain, use, and weigh information).[26] An individual's need for support and assistance to exercise these abilities introduces two other decision-making statuses. *Supported decision making status* 'distributes decision-making abilities required for competent decision-making processes across an individual and his or her supporters' but crucially, this is '*as directed* by the individual's will and/or intention'.[27] In Bach and Kerzner's words,

> Recognizing the role of support and representation in a supported decision-making process shifts the focus of competency from the individual, to the decision-making process. A competent decision-making process is one in which supporters and *representatives are guided by the will and/or intentions of the individuals* in ways that give the individual decision-making capability.[28]

[24] 'General Comment No. 1', para. 21.

[25] Gerald Quinn, 'Personhood and Legal Capacity: Perspectives on the Paradigm Shift of Article 12 CRPD', (paper presented at Conference on Disability and Legal Capacity under the CRPD, Harvard Law School, Boston, 20 February 2010), quoted in Bach and Kerzner, *A New Paradigm*, p. 86.

[26] Ibid., pp. 83–4. [27] Ibid., p. 24, emphasis added.

[28] Ibid., p. 87, emphasis added.

By contrast, *facilitated decision-making status* applies in temporary circumstances in which individuals' impairments prevent them from acting independently and they also lack supportive others who have personal knowledge to determine their will and preferences. This status permits some substituted decision-making based on the principle of best interests. Ultimately, however, the 'best interpretation of will and preferences' remains standard for these instances where substituted decision-making is required.[29] Even this status does not define them as 'legally incapable', nor is thought to reflect individuals' cognitive abilities or status.[30]

The CRPD has many welcome features, such as its normative focus on the social inclusion of and equal respect for individuals with impairments and outright resistance to outdated, harmful presumptions about their lesser status. It likewise provides an important critique of the supposed scientific evidence and value- and sociocultural neutrality underlying mental capacity as a concept that has traditionally sanctioned the paternalistic, discriminatory treatment of those with impairments. The Committee on the Rights of Persons with Disabilities correctly observes that '[t]he concept of mental capacity is highly controversial in and of itself' and 'is not, as is commonly presented, an objective, scientific and naturally occurring phenomenon' but 'contingent on social and political contexts, as are the disciplines, professions and practices which play a dominant role in assessing mental capacity'.[31] Moreover, the 'will and preferences' paradigm draws necessary attention to the process and justification behind substituted decision-making mechanisms in existing mental capacity regimes.

Yet the internal contradictions embedded within the prominent 'will and preferences interpretation of the CRPD weaken their critique of mental capacity, particularly when we consider how relational circumstances can impact positively or negatively on decisional abilities. Contradictory assumptions at the heart of the 'will and preferences' paradigm are exposed when we question whether safeguarding interventions are justifiable in cases where individuals with impairments are abused by their immediate support/care relationships. On one hand, the private sphere of individuals with impairments must be protected, including their right to make choices about their relationships, even if harmful to themselves. This is what it means to treat individuals with impairments *equally:* the logic is, if individuals *without* impairments can choose harmful relationships without outside intervention, then this should be

[29] Ibid., p. 91. [30] Ibid. [31] 'General Comment No. 1', para. 14.

no different for those *with* impairments. On the other hand, society and public institutions have positive obligations to ensure the promotion of individuals' legal capacity. Added to this, the CRPD expressly states in Article 16 that 'all appropriate legislative, administrative, social, educational and other measures to protect persons with disabilities both within and outside the home, from all forms of exploitation, violence and abuse'.[32] It further stipulates that there should be 'all appropriate measures to promote the physical, cognitive and psychological recovery, rehabilitation and social reintegration of persons with disabilities who become victims of any form of exploitation, violence, or abuse, including the provision of protection services.'[33] States are obliged to establish 'effective legislation and policies, to ensure that instances of exploitation, violence and abuse against persons with disabilities are identified, investigated and, where appropriate, prosecuted.'[34] In other words, the 'will and preferences' paradigm asserts the traditional liberal separation between private and public, yet recognition of the positive, supportive interventions that are needed to exercise these legal rights and capacity – and indeed, to comply with Article 16 – demands undercutting the very same private/public dichotomy. I explain this more fully in the next section.

III. The Private/Public Distinction

Within the strict reading of the 'will and preferences' paradigm, individuals with impairments are entitled to certain resources, goods, and supportive mechanisms necessary to exercise their legal capacity. This entitlement nonetheless does not translate into a positive duty to intervene in private, potentially harmful choices of the individual. To use a case study, imagine Emma has a combination of learning and cognitive impairments; she lives with her father who struggles to provide the necessary care and support Emma requires. He has a fractious, deteriorating relationship with social care workers, claiming their attempts to provide support are both unnecessary and invasive. This ongoing conflict confuses Emma herself: on one hand, she expresses her interest and preference for engaging in the activities offered to her by social care workers (such as attending college, going on outings), yet she is loyal to her father who claims these individuals are interfering with their lives. He forbids Emma from seeing these social care workers, increasingly isolates her from contact (i.e. prohibits her from answering the door or telephone by

[32] CRPD, Art. 16 (1). [33] Ibid., Art. 16 (4). [34] Ibid., Art. 16 (5).

punishing her if she disobeys, repeatedly tells her that he is the only one who can take care of her, that she cannot survive without him, that she must rely on him solely). There is a worry amongst social care workers that, should this enmeshment and isolation continue, the various skills she has acquired through outside contact will be suppressed, meaning her future autonomy might be at risk. Attempts to have outside mediation have been unsuccessful due to the obstruction of Emma's father. The few times social care workers have managed to speak to Emma, she tells them that she strongly desires to attend the various activities she had access to previously, as she enjoyed experiencing and learning new things. But when asked about whether she would like to live elsewhere apart from her father, she expresses that she 'wishes to stay with her father since she is completely helpless without him'.

These types of situations occur frequently when individuals with impairments choose to remain with carers who abuse their power to neglect, coerce, or abuse them, even against the supportive advice of others. Interpretations of Article 12 assert that the will and preference of Emma is legally binding should all else fail,[35] given that any substituted decision-making – and by default, third-party interventions – would violate Emma's legal capacity. Lurking behind this conclusion is an adherence to the liberal distinction between public and private spheres – partly due to how rights language typically asserts the protection of the individual and certain areas, such as family life, from outside intrusion. Or if such interventions are permitted, they must be strictly limited in accordance with the principle of proportionality.[36] Given how pervasively individuals with impairments have been imposed upon with some undesirable paternalistic treatment, most advocates of Article 12 welcome the move to defer to a person's expressed will and preferences and shore up the right of autonomy.[37]

Yet some circumspection about the liberal concept of individual rights is necessary, especially in light of how it constructs and shapes our notion of what is private and public. Though pervasive in the liberal imagination, conceptions of rights that rest implicitly on the public/private distinction ultimately determine contestable spheres of regulation and

[35] Cf. the case study discussed in Bach and Kerzner, *A New Paradigm*, pp. 144–5.

[36] See the analysis of Article 8 in Equality and Human Rights Commission, *Human Rights Review 2012; How fair is Britain? An assessment of how well public authorities protect human rights* (London: EHRC, 2012), pp. 263–84. Also Lacey, *Unspeakable Subjects*, p. 30.

[37] Dhanda, 'New Human Rights Lexicon', p. 45.

non-regulation.[38] I discuss this problem in more depth later. But even at the level of legal practice, the private/public distinction remains mythical. Regulation of the family is a case in point. Truancy laws in the United Kingdom is one mundane example where parents can be fined for excusing their child from school for family holidays during term-time.[39] If the private/public, non-regulated/regulated distinctions hold, parents would presumably have the right to make decisions which affect their children, including the right to decide when their family goes on holiday. But truancy laws and the authority of schools regulate if and when children can be legitimately excused.

For examples more pertinent to my argument, consider legal interpretation of Articles 3 and 8 in the European Convention of Human Rights (ECHR). Whilst the CRPD lacks the legal status of the ECHR, closer examination of ways that relational abuse has been addressed under the ECHR is instructive for my analytical purposes.[40] Discussing Article 3, Herring points out that the prohibition of torture and inhumane or degrading treatment or punishment is an absolute right, making it impermissible for this right to be infringed in any manner. The conventional public/private dichotomy is subsequently challenged, given that the state's duty to intervene overturns a family's right to privacy in cases of intimate abuse where one person's Article 3 (ECHR) rights are infringed.[41] Herring puts this point even more strongly, 'in an intimate abuse case the state

[38] Ibid., p. 30.

[39] See Press Association, 'More Parents in England Prosecuted for Taking Children out of School', *The Guardian*, 12 August 2015.

[40] There are notable differences between the ECHR and CRPD that we should keep in mind: unlike the CRPD, the ECHR (i) is a supranational statute with its own court; (ii) includes rights that are enforceable within contracting states (and thus, in English courts); and (iii) has its own jurisprudential developments and history. As Jill Stavert notes, the fact that state obligations can only be applied vis-à-vis ECHR rights has meant that recognition of positive duties of reasonable accommodation for those with impairments has been more gradual under ECHR law. That said, there are important ways that the CRPD could strengthen interpretations of ECHR rights. In *Glor v Switzerland* App. No. 13444/04, Chamber judgment of April 30, 2009, the ECHR referred explicitly to the CRPD in calling for the implementation of reasonable accommodation, as well as the need to protect individuals with impairments against discriminatory treatment (even as Switzerland at that time had not yet ratified the CRPD). Stalvert identifies further areas of potential overlap, such as rights under CRPD Art. 12 of the CRPD and ECHR Art. 8. In general terms ECHR rights could be interpreted more widely in support of CRPD rights to social inclusion. See Jill Stavert, 'UN Convention on the Rights of Persons with Disabilities: Possible Implications for Scotland for Persons with Mental Disorders', *Scottish Human Rights Journal* 47 (2009).

[41] Herring, *Caring and the Law*, pp. 282–3.

cannot justify its failure to protect a victim's Article 3 rights by referring to that person's right to respect private life.' Thus, positive obligations by the state can and do override private rights in Article 3.[42]

But what about cases where victims 'consent' to or are 'happy' with their mistreatment? Though these borderline cases are unlikely to qualify as instances of inhuman or degrading treatment under Article 3 of the ECHR,[43] their consideration under Article 8's protection of the right for private and family life likewise challenges the liberal private/public distinction. For example, Munby J (as he was then) sided with the local authority in removing a young man with severe learning impairments from the restrictive care of his father in the influential pre-MCA inherent jurisdiction case, *Sheffield City Council v S* [2002] EWHC 2278 (Fam). The judge stated that the father

> cannot pray Article 8 [ECHR] in aid as a trump card. On the contrary, and as Botta shows, the State, even in this sphere of relations between purely private individuals, may have positive obligations to adopt measures which will ensure effective respect for the son's private life. Thus the State, in the form of the local authority, may have a positive obligation to intervene, even at the risk of detriment to the father's family life, if such intervention is necessary to ensure respect for the son's Article 8 rights.[44]

Munby further cited Sedley LJ in *Re F (Adult)* referring at p 57E to Article 5 of the ECHR:

> The family life for which Article 8 [ECHR] requires respect is not a proprietary right vested in either parent or child: it is as much an interest of society as of individual family members, and its principal purpose, at least where there are children, must be the safety and welfare of the child. It needs to be remembered that the tabulated right is not to family life as such but to respect for it. The purpose, in my view, is to assure within proper limits the entitlement of individuals to the benefit of what is benign and positive in family law. It is not to allow other individuals, however closely related and well-intentioned, to create or perpetuate situations which jeopardise their welfare.[45]

[42] Article 3 infringements could also be interpreted as an issue of conflicting individual rights – the right of protection (which generates positive state obligations) and the right of autonomy/privacy. This nonetheless does not affect the claim that the positive duties incurred as a result the individual's right of protection can override the negative right to privacy.

[43] Ibid., 283.

[44] *Sheffield City Council v S* [2002] EWHC 2278 (Fam), para. 39.

[45] Ibid., para. 43.

Two points are notable in Munby J's interpretation of Article 8 (ECHR): first, the right to private and family life is a qualified right that must be balanced against another individual's Article 8 right. As such, it is a balance between two competing negative rights. But the second aspect goes further: the positive obligation of the state to intervene in situations where the family fails to protect the bodily and psychological integrity of the individual (in other words, when Article 8 paragraph 2 is satisfied) *qualifies* this right to private and family life.[46] Where competing negative rights of private individuals and positive obligation of the state require balancing, Herring argues that the law must consider the *values* that underline these rights.

> In the case of Article 8 the underlying value is that of autonomy: the right to pursue your vision of the 'good life'. A judge could then consider the extent to which the proposed order would constitute a blight on each of the party's opportunities to live the good life and make the order which causes the least blight. Applying that in this context I would argue that although removing the victim from intimate abuse from an abusive carer will infringe the carer's autonomy, it will do so to a much lesser extent than leaving the victim to suffer abuse would do. But what if the victim does not want the assistance? Here there is a balance between protecting the current autonomous wish of the victim, with the increase in autonomy they may experience if they were removed from the abuse.[47]

According to this analysis, the application of Article 8 of the ECHR has to transcend the public/private dichotomy in order to fulfil requisite obligations in certain situations. This may very well mean intervening on privately chosen, but abusive or disabling, relationships, particularly if this enables the individual to develop and exercise her autonomous agency more fully.[48] Private relationships characterised by abuse, manipulation, or coercion inhibit an individual's potential to develop autonomy skills, and, as we will see more fully in Chapter 5, the failure to intervene can itself signal the neglect of positive duties of support to encourage the autonomy of those with impairments.[49]

[46] ECHR, Art. 8, para. 2: 'There shall be no interference by a public authority with the exercise of this right except such as is in accordance with the law and is necessary in a democratic society in the interests of national security, public safety or the economic well-being of the country, for the prevention of disorder or crime, for the protection of health or morals, or for the protection of the rights and freedoms of others.' Also see Herring, *Caring and the Law*, p. 283.

[47] Herring, *Caring and the Law*, p. 284.

[48] I expand on this point in Chapter 5.

[49] The next chapter discusses 'autonomy competencies' in more depth.

One could argue that the flexible interpretation of the CRPD's Article 12 mitigates this concern. Bach and Kerzner acknowledge explicitly the widespread abuse and coercion of individuals with impairments and recommend safeguarding mechanisms to protect individuals from 'serious adverse effects', particularly those who are isolated, with limited financial resources. 'Serious adverse effects' occur when individuals experience (a) loss of property or necessities for themselves and their dependents; (b) serious illness, injury, or are deprived of liberty and personal security; or (c) threats or attempted threats to cause physical/psychological harm to oneself; violent or threatening behaviour that generates cause for others to fear physical/psychological harm from oneself.[50] Various representatives, facilitators, and monitors will evaluate what is required to protect the individual's legal capacity in cases of serious adverse effects and provide supports to the individual in question. For example, a Legal Capacity and Support Officer will be assigned to 'arrange supports as needed to address situations where serious adverse effects are occurring or may occur and there is reason to believe that a person's ability to make and/or act on their decisions will be enhanced by such supports'.[51] An Administrative Tribunal would adjudicate disputes around appropriate supports for the individual.

On one hand, the establishment of these various monitors (the Legal Capacity and Support Officer, the Administrative Tribunal) suggests that in certain situations of severe adverse effects, and depending on the decision-making status of the individual, there may be a positive duty to intervene *even if* the individual in question opposes it. As they state:

> The Legal Capacity and Support office would have the authority to investigate concerns, complaints, and allegations of serious adverse effects in situations where individuals are in a supported or facilitated decision-making status, or where there are reasonable grounds to indicate that a person is unable to act legally independently.[52]

In other words, public bodies have an obligation to investigate suspicions of abuse involving individuals with a particular decision-making status, even as he or she might insist on staying with their abuser. It would follow that the safeguarding powers of monitors may well recommend overruling the individual's preference to remain in an abusive situation in order to best support her legal capacity. But Bach and Kerzner eventually reject

[50] Bach and Kerzner, *A New Paradigm*, pp. 174–5.
[51] Ibid., p. 120. [52] Ibid., p. 141.

this conclusion, arguing alongside the strict interpretation instead that the rights, will and preferences of the person has priority if we are to protect individuals with impairments from a paternalistic approach to safeguarding. Positive obligations to protect individuals from abuse, if finely balanced, should defer to respect for one's autonomy so that 'people have the legal capacity to say "no" to others who would impose treatment or confinement, or a particular service upon them in the name of protection'.[53] In contrast to a protectionist approach, respect for individuals' legal capacity will outweigh future interests in physical, emotional, psychological safety or long-term autonomy. From the standpoint of both strict and flexible interpretations of Article 12, the possibility that the safety and long-term autonomous agency of individuals may not always be secured may be a price worth paying in order to respect their expressed choices (which is coextensive with respect for their legal capacity).

Issues of safeguarding are relatively straightforward if the individual in question is receptive or amenable to outside support. But legal cases evidence time and time again that those who are in such relational circumstances frequently resist outside intervention and demand to remain with their abusers who continue to disable their agency[54] or, indeed, choose to live in circumstances of poor care or self-neglect, deeply suspicious of other caring relations or outside agencies who want the person to receive better, enabling care.[55] Bach and Kerzner's approach fails to properly address the issue of what is justifiable to do (or not do) in these scenarios.[56] It is a relatively optimistic view that individuals and those with whom they are in relationship will welcome supports offered by monitors like the Legal Capacity and Support Officer or would be willing to attend an Administrative Tribunal; in situations of abuse, coercion, exploitation – and indeed, where impairment might cause misunderstanding of their unsafe physical circumstances – hostility towards,

[53] Ibid., p. 95.

[54] *A Local Authority X v MM & Anor* (No. 1) [2007] EWHC 2003 (Fam); *A Local Authority v WMA & Ors* [2013] EWHC 2580 (COP), *A Primary Care Trust v P & Ors* [2009] EW Misc 10 (EWCOP); *A Local Authority v DL* (2011) EWHC 1022 (Fam); *A Local Authority v A & Anor* [2010] EWHC 1549 (Fam).

[55] As is common in cases of dementia. My thanks to Tony Hope for raising this important example where individuals might be in disabling circumstances but are surrounded by relationships that do genuinely care for them and want the best for their well-being.

[56] This indicates a default emphasis on formal equality rather than substantive equality, given that individuals with impairments often have diminished capacity on their own, where their reliance on others could heighten their susceptibility to abuse in certain environments. I discuss this more fully in Chapter 5.

and suspicion of, outsiders remains a real possibility. Would this support then be imposed on the individual? Sometimes staying in an abusive, neglectful but familiar situation will seem the more natural option as opposed to leaving to go into a wholly unfamiliar, frightening, (albeit) supportive context. If this is true, then we need to question the ways in which adherence to the private/public dichotomy reinforce those relationships that compromise the autonomous agency of individuals with impairments. The inquiry into what our obligations and duties are in terms of intervention *does not end* when individuals actively resist support and choose to remain within abusive or simply disabling situations. If anything, we need to explore and establish normative parameters for justifiable interventions in such cases.[57]

III. Negative/Positive Liberty and the Assumption of Individualism

The public/private dichotomy may persist in interpretations of the CRPD's Article 12 due to its commitment to an underlying picture of the autonomous rights-bearer that is oddly individualistic, where the expression of liberty revolves around the protection of one's choices from outside incursions. Initially, Bach and Kerzner appear to argue that liberal autonomy has tended to be overly individualistic, focusing more on negative liberty. The right of autonomy thus acquires an inflection that disregards the importance of relationships and environmental supports for the ability to exercise one's autonomy. Here I agree, and the next chapter mounts a similar challenge.

But grounding Bach and Kerzner's approach is a slightly incoherent resolution to Isaiah Berlin's negative/positive liberty dichotomy. Berlin envisages negative and positive liberty as mutually exclusive concepts. He famously defends negative liberty as an important protection of individuality against potential state intrusions or restrictions sanctioned by accounts of positive liberty, on grounds that individuals can be 'forced to be free'. The problem for Bach and Kerzner is how to frame positive obligations of the state to assist individuals in exercising their legal capacity, yet ensure their personal autonomy is not unduly restricted at the same time. Contra Berlin, they suggest that negative and positive liberty approaches to autonomy are 'entirely interdependent' where, the former is vital to 'ground citizens' rights to refuse interventions by others' whilst 'positive obligations of the state . . . ensure people have access to

[57] I discuss this issue explicitly in Chapter 5.

supports and capabilities to actively exercise their autonomy'.[58] Adopting the analogy of Gerald Quinn, Bach and Kerzner state:

> [L]egal capacity is both a 'sword' to advance positive freedom and make one's way through the world in 'un-coerced' relations with others; and a 'shield' protecting against others who would impose decisions upon you.[59]

If I understand this correctly, their argument (alongside Quinn) is that legal capacity functions as a protective sphere against outside interference into one's private choices, whilst expressing one's entitlement to certain goods and resources. In this way, Bach and Kerzner argue for the compatibility of positive and negative freedoms in legal capacity.

However, this rather odd formulation rests on some conceptual confusion about the precise relationship between positive and negative freedom. Both forms of liberty are linked, but not in the manner that Bach and Kerzner suggest. Interpretations of Article 12 seek to stress the illegitimacy of third-party intrusions, to emphasise how respect for an individual's choices should outweigh welfarist, paternalistic concerns. But this articulation sits uneasily with the concept of positive liberty. In its most basic form, positive liberty emphasises that freedom is an expression of how the will is *structured* and *motivated*; it requires value judgements about one's ends, desires, and preferences. According to this formulation, interventions by others can be justifiable. Or to put it differently, the premises of positive liberty do not immediately rule out others imposing decisions upon you.

Why this is so is due to the incoherence at the heart of negative liberty, thought of purely as freedom from restraint or an *opportunity-based* concept. As Charles Taylor has convincingly shown, opportunity-based concepts of freedom, premised either on the absence of constraints, or indeed the lack of initial positive supports one needs to act in the world without interference, is untenable without presupposing freedom as an *exercise concept* – where we analyse the *quality* of one's agency and what actually *counts* as obstacle to freedom. This means that even when we purport to advance a stance of negative freedom (where we stress freedom from interference), we necessarily appeal to the type of substantive evaluation that is more characteristic of positive liberty.[60] To speak of freedom from restraint purely in the sense of external restraints,

[58] Bach and Kerzner, *A New Paradigm*, p. 42. [59] Ibid., p. 43.

[60] Charles Taylor, 'What's Wrong with Negative Liberty?' in *Philosophy and the Human Sciences: Philosophical Papers 2* (Cambridge: Cambridge University Press, 1985), pp. 211–29.

of protection against outside intrusions and impositions, does nothing to address the *internal* restraints and barriers that can restrict one's freedom. Prioritising the internal domain of freedom ignores how, even there, unfreedom can occur. If unfreedom occurs *internally*, then it is question-begging as to why we would say external barriers somehow violate our freedom more than those internal barriers that directly impede the functioning of our motivations or knowledge of our authentic wishes.[61]

The 'will and preferences' interpretation assumes that the positive supports and obligations provided by public institutions will prioritise and adhere to negative liberty as an opportunity concept, to liberal commitments of value-neutrality. In reality, determining what counts as constraint invokes certain judgements that violate those very commitments; resolving the impasse of *which* restraints are acceptable or not demands some comment, discussion, or judgement about the type or source of motivation that one claims as genuinely free or unfree. We have already moved beyond the remit of negative liberty at this level of analysis. Contra Bach and Kerzner, the interplay between negative and positive liberty fundamentally *questions*, rather than asserts, the priority of the subjective domain. This is why it is important to resist the presumption that the concepts of liberty and freedom are coextensive with the concept of autonomy. Liberty articulates the *conditions of action*, whereas autonomy articulates the *conditions of willing*. Liberty is a necessary but *not sufficient* condition for autonomy. I don't wish to get too bogged down by the philosophical argument – the next chapter discusses autonomy in detail. But this slippage between concepts, whilst implicit within the Committee's General Comment, occurs explicitly within Bach and Kerzner's analysis.[62] Assuming that Article 12 of the CRPD concerns the right of autonomy, appeal to the positive/negative liberty distinction seems both imprecise and misguided.

[61] One could argue that this debate ultimately revolves around what constitutes freedom. Even if it is the case that negative freedom implies positive freedom, proponents of negative freedom might nonetheless assume that there is an individual 'will' that is separable from external influences or forces, and that its protection from such forces is valuable. However, I ultimately find this rejoinder unsatisfactory. The key point Taylor makes, as I understand, is that this position still does not answer the question of what happens when the internal will experiences compulsion of some sort (i.e. addiction, emotional dysregulation). The intuition that there is an essential, pure will to protect from outside intrusion does nothing to track the heteronomy that is inherent to such cases of pathological compulsions. My thanks to Alice Obrecht for raising this objection.

[62] Bach and Kerzner, *A New Paradigm*, pp. 38–44.

Let me nonetheless assume for the sake argument that liberty and autonomy are intrinsically connected (or possibly equivalent) in Article 12. Even then, Bach and Kerzner's contradictory use of the positive/negative liberty distinction is symptomatic of a rather counter-intuitive commitment to liberal individualism. That freedom amounts to 'making one's way through the world in uncoerced relations with others, as a shield protecting one against others' reflects an oddly atomistic view, where the right of autonomy expresses itself through the assertion of one's subjective will and choice without interference. Interpretations of Article 12 do claim that autonomy is a *relational* concept[63]; however, assumptions of relationality are used mainly to minimise *differences* between individuals with and without impairments – to assert that all are embodied individuals, dependent on others at some point in our lives, who require the care and assistance provided by relationships and social, public goods. It is to make space for the requisite positive obligations of support needed for individuals to realise their legal capacity. At the same time, however, the locus of decision-making – the source of our motivation, preferences, values – remains subjective and individual. The expressed 'will and preferences' of a person is assumed as authentic; it is the ultimate trump to safeguard against outside incursions into one's private life.

Autonomy as a negative liberty, shield concept exercises a powerful influence in the liberal understanding of individual rights. The problem with the 'wills and preferences' interpretation is that it buys into the idea that the point of rights is to protect the bounded individual making decisions on their own; it reduces the content of the right of autonomy to a negative liberty, opportunity-concept, coextensive with the expression of one's subjective preferences. But these reductive assumptions fail to capture the full complexity of how socialisation influences the development of personal autonomy.[64] Moreover, they misunderstand how law and legal rights structure relationships in fundamental ways. As we will see in the next section, a relational approach to rights exposes fully the implicit individualistic commitments of the 'will and preferences' interpretation of legal capacity. Relational rights will further help lay crucial groundwork for establishing a more dialogical, relational concept of mental capacity.

[63] Ibid., pp. 40, 84.

[64] I have much more to say about this in my discussion of relational autonomy in the next chapter.

IV. Relational Rights

In the preceding sections, I argued why the concept of legal capacity struggles to displace the concept of mental capacity, even as its critique of the latter may have some merit. The 'will and preferences' paradigm rightly draws attention to some problematic dimensions surrounding mental capacity but is ultimately unsuccessful in arguing for the complete disposal of the concept, particularly in light of its weak response to ways in which the relational context can compromise an individual's decision-making ability. As discussed so far, the 'will and preferences' paradigm contains questionable assumptions that mean (i) the pre-eminence of subjective preferences, regardless of how these might reflect systemic, relational abuse; and (ii) a failure to see that positive obligations and rights do not cohere easily with a negative shield concept of the right of autonomy. To be clear, I am not making an argument about the *grounds on which*, or *the nature of* justifiable third-party interventions. I explore these issues explicitly in Chapter 5.

What I am arguing is that we need a concept of mental capacity that can help capture the relational dimensions that enable or disable an individual's decisional agency. Given that mental capacity is thought to safeguard the right of autonomy, we need to establish how the law and its application of rights already presuppose relationality; the law structures relationships in particular – and sometimes contestable – ways. Such an analysis requires resisting an influential cluster of liberal ideas underlying current interpretations of legal rights. Rights are typically conceived of as an intrinsic property and entitlement of individuals, constructed to protect the bounded, rational individual or the private domain from the intrusions of the public or political sphere. Or to use Dworkin's words, rights are premised on formal equality so that individuals are 'entitled to the same concern as others' and function as 'trumps' for individuals to advance or enforce their interests against the competing interests of others or the political state.[65] The CRPD and 'will and preferences' paradigm are not immune to these liberal assumptions that likewise exercise a pervasive influence in the common law focus on the individual as rights-bearer. These assumptions ignore not only ways in which rights can be constitutively realised through the assistance others, but also how

[65] Ronald Dworkin, 'Rights as Trumps', in Aileen Kavanagh and John Oberdiek, eds., *Arguing about the Law* (Abingdon: Routledge, 2009), pp. 335, 344.

interpretations of rights may structure and sanction relationships that fundamentally disable or exclude individuals.

Feminist legal scholars and philosophers pick up on these two points in their challenge to mainstream liberal approaches to rights. On one hand, rights discourse in the abstract articulates underlying values about the dignity and equality of all persons; on the other hand, its traditional usage contains questionable assumptions that can impede the substantive promotion and protection of these values.[66] These critiques do not reject wholesale the language of rights, but reform its underlying assumptions along relational lines. A relational account of rights makes the following claims: (i) the rights-bearer as *embodied self* is nested within relationships and sociocultural contexts; (ii) legal rights reflect *value-laden ways of structuring relationships* that affect individuals' enjoyment of core values. In making these claims, a relational approach critically scrutinises the problematic consequences of these structures, bringing to the forefront debate about contesting values and their complex actualisation through legal rights.

First, relational approaches to rights argue that individuals are embodied and interdependent as opposed to the bounded, rational, solitary picture of the liberal self. Features of our bodily selves draw attention to the fact that we are prone to certain abilities and vulnerabilities that rights must accommodate. Recognising embodiment allows us to foster appropriate legal responses to ways the body can limit the exercise of rights, particularly the right of autonomy. According to Nedelsky,

> This sort of attention to the body allows a recognition that the body itself is a source of knowledge for this process. Reflecting on the limitations of the body, and what it means to optimally interact with those limitations, has the potential to reveal the importance of a sympathetic response to limitations on autonomy. . . . It is important that the concept of autonomy embraces an image of the embodied self, or we will not be able to adequately explore what fosters reason and autonomy.[67]

The exercise and protection of our rights will imply a network of support. The interaction between our relational context and our individual bodily reality shapes certain skills necessary for decisional and practical agency. Or to put the point negatively, contexts that deride or doubt, or

[66] Dworkin's distinction between the abstract and concrete is helpful here. See *Taking Rights Seriously*, p. 93.

[67] Nedelsky, *Law's Relations*, p. 172.

institutions that systemically exclude, can be directly implicated in the incapacity of one's agency. This recognition is also important because we become more attentive to differences between individuals and its impact on dynamics of power and status:

> By excising the body and affect from the essence of the rights-bearing self, the multiplicity of differences among people is removed as well. Conversely, when the conceptions of reason and autonomy have the body and affect integrated into them, the differences that both make manifest become central. The realities of differences in abilities and in emotional states – as well as the relational differences of power and status – are no longer presumed to be marginal to the issue of equal rights; they appear as integral to the full particularity of the subject of those rights.[68]

On one hand, we could say that negative rights – for example, against torture – do in fact try to protect bodily integrity. But even at that level, rights protection essentially revolves around the *abstract* body – the body that is the subject of universal, formal equality. By contrast, Nedelsky's argument here asserts the particularistic nature of how one's embodiment is uniquely expressed, manifests itself, and is reinterpreted in our social and legal context. For example, consider the abusive exploitation of residual difficulties intrinsic to certain bodily impairments.[69] At root, we are all vulnerable through the reality of our physical interests, needs, and limitations. But this inherent vulnerability can be exploited more or less, in certain situations and with certain practices, depending on the extent to which we rely on others to cope with the daily realities of our embodiment.[70] Appreciating how an individual's particular impairments can make them susceptible to limited opportunity or harm from others is not disablist – indeed, such contextual sensitivity is vital in order to fulfil the positive obligations owed to the individual, to secure and promote her own potential for decision-making and action in light of her unique, bodily reality. This claim does not essentialise or reduce individuals to the level of the body, but challenges assumptions that the *subject* of rights is either the disembodied, rational subject, or an abstract, indistinguishable body that is universally protected. Individuals with impairments are

68 Ibid.

69 I realise the notion of residual difficulties inherent to certain impairments is controversial according to the social model of disability. I explore this view more fully in Chapter 5.

70 Here I draw upon Catriona Mackenzie's threefold distinction of vulnerability in her paper, 'The Importance of Relational Autonomy and Capabilities for an Ethics of Vulnerability,' in Catriona Mackenzie et al., eds., *Vulnerability: New Essays in Ethics and Feminist Philosophy* (Oxford: Oxford University Press, 2014), p. 47.

therefore more likely to be perceived as substantively equal, not diminished, subjects of rights.[71]

Second, a relational analysis of rights challenges value-neutral, descriptive attributions of law, examining instead ways that the law reflects particular value-laden interpretations that structure relationships or perpetuate unequal power accordingly. This fundamentally contests the abstracting interpretive move of liberal rights. Abstracting rights-bearers from the sociocultural context, for example, obscures ways that social norms of gender, culture, and religion, as well as socioeconomic status (i.e. wealth or poverty) can be important constituents of personal identity *as well as* relations and misuses of power. Strong egalitarian reasons may underlie the historical focus on the abstracted, decontextualised individual as the bearer of rights. However, the mask of formal equality can shield from immediate view how application of the same legal rights can lead to discriminatory, exclusionary, and disempowering consequences, embedding ideological power and systemically excluding or oppressing certain groups.[72]

On these grounds, prevailing descriptive analogies of liberal rights can be found to be deeply problematic. Most prominent amongst these is the private/public, non-regulate/regulated boundary. This boundary initially appears as a value-neutral characterisation of areas where the law cannot intervene. The private sphere is often described as descriptive fact or a feature of biology – the individual is literally a bounded self, the family is a bounded unit, the house is a bounded structure – yet history indicates that non-regulated areas of society are political, legal constructions that frequently legitimate contestable norms and inequitable practices. Feminists argue that these descriptions reflect complicity towards certain existing social arrangements and power relations that run through the private sphere.[73] As Nicola Lacey puts it, 'non-regulated areas may be seen as areas in which the legal system implicitly legitimises sexism and racism, given the social facts of their existence.'[74]

Consider, for example, how the private/public division of liberal rights shape questionable responses to violence against women.[75] According to

[71] Nedelsky, *Law's Relations*, p. 191.

[72] Private property rights, for example, have been the historical preserve of men of a certain class, to the disenfranchisement of women, indigenous groups, and the disabled. See Lacey, *Unspeakable Subjects*.

[73] Ibid., p. 29. [74] Ibid., p. 30.

[75] This is not to say that I ignore the various incarnations of domestic battering or intimate abuse (i.e. men abused by women, women abused by women, men abused by men). I use

Nedelsky, the law faces a strange conundrum in cases where women kill their battering partners: on one hand, feminist defense lawyers emphasise expert testimony that provides evidence of abuse, violence, and its resultant psychological damage – what is commonly called 'battered women's syndrome'.[76] This syndrome helps capture ways that relational abuse fundamentally inhibits and damages the skills and sense of self that are required to exercise individual autonomy. It also mitigates the assumption that the abused could leave if the abuse was so severe. On the other hand, battered women's syndrome reinforces simplistic, harmful gender stereotypes – such as the helplessness, passive victimhood, and non-agency of women – that contribute to ongoing inequality and patriarchal oppression. Thus, women who kill abusive partners out of self-defence are characterised as simultaneously reasonable and incapacitous.

But whereas the law currently focuses more on what this contradiction means in terms of the culpability of the individual agent, Nedelsky shows that a *relational* analysis would concentrate on how relationships and social environment impact on her agency. A woman who repeatedly returns to a battering partner can be a reasonable agent, yet particular relationships compromise and damage her autonomy.[77] Going further, Nedelsky highlights that battered women are kept in abusive situations, not just due to their private relationships. Societal failures to protect women also function as coercive forces that keep battered women in these situations, since the absence of societal options make exit unavailable even when she faces death threats.[78] In Nedelsky's words:

> It would seem a full understanding of the impact of battering on a woman's autonomy requires attention both to deep psychological impairment – feelings of worthlessness, dependence on the batterer, difficulty in seeing a way out, a profound sense of helplessness – *and* the sorts of coercive force that, absent societal protection, even a fairly convention of autonomy would see as a serious constraint.[79]

In other words, societal, legal, public institutions that adhere to the private/public dichotomy are often disinclined to protect and intervene in

this common feminist example mainly to illustrate my broader point here that the law already makes particular judgements about the boundary between private relationships and public intervention, about the types of relationships that are desirable (or warrant intervention), all of which can lead to differential costs on individuals in terms of their substantive equality and freedom.

[76] Nedelsky, *Law's Relations*, pp. 175–83.

[77] Ibid., p. 176. [78] Ibid., p. 181. [79] Ibid.

individuals' chosen relationships. It is deemed a private matter: the battered woman will make her own choices, even as this raises contradictions in how we characterise her agency (as reasonable-but-not-autonomous). Nedelsky argues that to capture the full scope and depth of harm to women requires a fundamental shift in understanding, away from the private/public dichotomy. Battering is not just a private affair that happens between individuals, but a *social phenomenon* that persists through the systemic patterns of behaviour by various levels of public officials (police, prosecutors, judges) and society (neighbours, friends, and family), all of which collectively *fail to protect women from intimate abuse.*[80] Conventional views about the private and public domains fall away as a result, as societal structures' failure to protect women are themselves implicated in domestic abuse. Failure to protect can occur in two ways: through the absence of external support, positive options, and assistance, as well as the *failure to sanction external interventions within certain limits.*

It may be that feminist jurists and philosophers are more open to the prospect of outside interventions to protect, given how the oppression of women has frequently occurred in what has been deemed the 'private sphere', where law and public policy have implicitly sanctioned practices of gender inequality. By contrast, the state's disrespect for the private person, through forced treatment and care, characterises the historical mistreatment of those with impairments. The move to shore up the 'private' is understandable given this context. Nonetheless, careful recognition of the historical misuse of public power is a necessary, *not sufficient*, condition to mitigate the abuse and mistreatment of individuals with impairments, which often goes undetected in the private sphere. We can and should debate the appropriate boundaries for intervention. That is my focus in Chapter 5.

But the important point for my present analysis is that, regardless of whether *or not* interventions are made in situations of domestic abuse, the law is structuring relationships in crucial ways between intimates, between the individual and what is typically demarcated as the 'public sphere'. Crucially, this structuring occurs even when private decisions are deemed legally sacrosanct. Acceptance of conventional boundaries between the private and public, non-regulated and regulated, is essentially to take a value-laden stance of what power relations are legitimate and valid, and – by implication – sanctions violence against women so long as it remains cosseted away in the private sphere. The language of rights already embeds

[80] Ibid., p. 183.

some discussion of values, of certain views about gender, as well as *the types of relationships that are desirable to the law and state* (e.g. whether same-sex couples are accorded the same legal protections as heterosexual couples; whether co-habiting relationships have the same legal protections as marriage). These may not be *ethically* desirable. Depending on the consequences for individuals, the law's structuring of relationships may fall short in realising common values – such as equality – or evince the differential, unequal impact of rights – meaning further critical reflection on these rights and values is needed to structure normatively justifiable relationships.

A relational analysis of rights in the case of domestic abuse suggests that one must accept that the private sphere has to become more permeable in order to mitigate oppressive, abusive relational practices. I would argue, alongside other feminist philosophers and jurists, that this logically follows from a relational analysis of rights that challenges the private/public dichotomy. As Nedelsky states, 'many "interventions" are in fact the removal of unjustified impunity for actions to which the state would otherwise react'.[81] She suggests that in the case of domestic abuse against women and children, a woman's responsibility needs to be contextualised within 'the consequences of both her personal relationship and how that relationship is situated in the wider structure of social and governmental relationships that effectively tolerate the abuse of children as well as women'.[82] Questioning the regulatory boundary between private and private could sanction moves similar to post-apartheid South Africa, where constitutional reform was undertaken, transforming the remit of judicial review in order to tackle discrimination in the 'private' sphere.[83] Commenting on the fluid private/public nature of sexual practices, Lacey similarly argues:

> Feminist arguments are perfectly consistent with the idea that sexual practices are among those from which people have the right to exclude others and the state, but point out that the range of seriously autonomy-reducing sexual practices which call for political critique and, sometimes, action go beyond those, such as rape, traditionally acknowledged to be harmful.[84]

In other words, a relational theory of rights must accept the consequences of their critical challenge to conventional boundaries between state and

[81] Nedelsky, *Law's Relations*, p. 72. [82] Ibid., p. 302.
[83] Ibid., p. 214. [84] Lacey, *Unspeakable Subjects*, p. 96.

society, public and private, regulated and non-regulated. Broader intervention into traditionally protected spheres is accepted as necessary to directly challenge violence against women. These boundaries are fluid and, as Lacey's example demonstrates, what is thought of as private can easily bleed into the public sphere, depending on how practices impact on the autonomy of the individual.

A relational account of rights may therefore gesture towards an opposite direction to that of the 'will and preferences' paradigm in an important way. Recall that the latter implies that third-party interventions into the lives of individuals with impairments, even in disabling relational practices, represents a fundamental violation of individuals' legal capacity.[85] In the first instance, both strict and rigid readings would likely concur with feminist relational analyses that stress the need to reform societal structures to improve provisions of support, exit options, and resources. However, where they part ways is perhaps in Nedelsky's claim that protective mechanisms need to be more robust. This conclusion is consistent with a relational analysis of rights that undercuts the private/public distinction at its root. By contrast, interpretations of the 'will and preferences' paradigm tend to gloss over the need for protective safeguards, which may sanction legitimate interventions in a person's abusive relationships. Resistance to this conclusion is understandable: after such historical, paternalistic mistreatment by public institutions, individuals with impairments do and should have the 'dignity of risk', to make mistakes and unwise decisions about their health and care treatments, living arrangements, and relationships. In no way am I denying that this is a valid proposition. But present interpretations of Article 12 have tended to minimise how avenues to protect individuals likewise need to accompany supportive structures. The Committee on the Rights of Persons with Disabilities recognises that these disabling relationships 'may be exacerbated for those who rely on the support of others to make decisions'. Whilst it is stressed that safeguards should be in place, particularly in cases where 'the interaction between the support person and the person being supported

[85] Michelle Madden Dempsey in *Prosecuting Domestic Violence; A Philosophical Analysis* (Oxford: Oxford University Press, 2009) suggests that the state's duty to intervene in domestic abuse is partly aimed at challenging patriarchy. In drawing analogies with feminist arguments, I am not suggesting in the first instance that the state's obligation to intervene in cases where individuals with impairments are abused is also part of a broader strategy to combat systemic disablism and discrimination. Although intuitively plausible, further argumentation would be required to make this stronger point and would extend beyond the scope of this chapter.

includes signs of fear, aggression, threat, deception or manipulation', they continue that such protection nonetheless 'must respect the rights, will and preferences of the person, including the right to take risks and make mistakes'.[86]

However, this response fails to see that protection is not always contrary to respect for the autonomy of individuals. Counterintuitive consequences result if we take traditional liberal rights, grounded in abstracting contextual factors and treating everyone the same in a formal sense. Safeguarding interventions in abusive situations between individuals with impairments and their carers or family members – particularly if the person concerned is 'happy' with this relationship – would be perceived as discriminatory, evidencing unequal respect for disabled individuals. Extending this logic, then, interventions in practices of rape or intimate abuse against those who choose to remain with threatening partners would likewise be discriminatory and unequal. On the face of it, the example of rape might simply reiterate a difference between what the law deems as 'illegal', public action (i.e. rape), compared with what is technically still 'legal', private, although ethically dubious (i.e. abuse, coercion of individuals with impairments, the case of Emma).[87] We might say that how rape is treated in the law merely recalibrates the conventional boundaries between public/private. Yet criminal prosecution of rape suggests a far more elastic boundary in the law, as interventions can be carried out *independently* of the victim's support for prosecution, precisely out of public interest.[88] The law itself tries to resist sanctioning instances of domestic battering and rape in certain circumstances, exposing how adherence to traditional liberal boundaries can *perpetuate* inequality and abuses of power. Such boundaries can sanction relationships that compromise equality and autonomy in a substantive, foundational way rather than protect and enable the agency of individuals who are already (or have been) systemically discriminated against.

Finally, a relational analysis of rights is amenable to exploring and evaluating the underlying values of rights and their ethical justification. Nedelsky states,

> A relational approach to rights calls for a constant inquiry into the values that are at stake in any given problem (e.g. of interpreting, implementing,

[86] 'General Comment No. 1', para 22.

[87] Thanks to Tony Hope for bringing up this point.

[88] 'CPS Policy for Prosecuting Cases of Rape', available online: https://www.cps.gov.uk/publications/prosecution/rape.html#05, accessed 1 December 2016.

> recognizing new rights). It then requires an engagement with the inevitably difficult, sometimes even speculative, process of figuring out what kinds of relations would foster those values and then, again, what kind of interpretation, policy of enforcement, or legal recognition would structure relations one way rather than another.[89]

The contested area of mental capacity law demonstrates the need for such analytical tools. Consider how we might interpret provisions of the CRPD. The CRPD, like the MCA, recognises the importance of personal autonomy. Yet its content, the types of legal and moral obligations that result, or even its assumed priority over other values, are all heavily contested between and internal to different legal instruments. Indeed, these remain controversial even when we set aside issues of impairment: for many, autonomy as a value is considered only one amongst many other incommensurable human goods.[90] Elsewhere in the CRPD goods other than autonomy are likewise emphasised: Article 24, for example, focuses on the right to education so as to promote individuals' 'human potential and sense of dignity and self-worth', along with their 'personality, talents, and creativity, as well as their mental and physical abilities'.[91] Echoing Article 16 of the CRPD, the Committee on the Rights of Persons with Disabilities further stress rights such as 'freedom from abuse and ill-treatment'.[92] From a relational analysis, rights related to freedom of abuse do not automatically gesture towards the negative liberty focus implicit within the 'will and preferences' paradigm. The rights associated with autonomy can and often do conflict with the rights associated with freedom from abuse and ill-treatment, or other positive rights that incur obligations of the state, such as the right to education, health care, and a minimum standard of living. On one reading, the promotion of education and freedom from abuse could overturn an individual's expressed rights, will, and preferences to remain in an inherently disabling relationship. Indeed, as we will see in subsequent chapters, positive action could be justified to provide basic constituent skills necessary for autonomy and practical agency.

Much of this likewise applies to the MCA. A relational analysis could pinpoint a number of important incongruous values within the statute. One prominent example is its conflicting conceptions of autonomous

[89] Nedelsky, *Law's Relations*, p. 343.

[90] Cf. Virginia Held, *The Ethics of Care: Personal, Political, and Global* (Oxford: Oxford University Press, 2005).

[91] CRPD Art. 24 1. (a), (b), (c).

[92] 'General Comment No. 1', para. 29.

agency. The functional test, for example, depicts a time-slice view of decision-making, removed from a person's broader life context or long-term values. Respect for autonomy in the functional test is coextensive with deference to an individual's subjective preferences in those particular circumstances. Best interests decisions, by contrast, assume diachronic, long-term agency, premised on consistent, overarching values and goals that can sometimes be determinative. The substitute decision-maker is directed to consider the 'person's past and present wishes and feelings' as well as the person's 'beliefs and values'.[93] Thus, even when one lacks capacity, an individual's right of autonomy could still be respected when her beliefs and values are given due consideration. The incommensurability of these interpretations of agency is apparent: should one prioritise an individual's present wishes at the expense of long-term values prior to capacity? Should we be focusing purely on an individual's present wishes, even as it leads to harmful, disabling consequences for her long-term decision-making abilities, or indeed, are contrary to her long-term values and wishes?

Thus, both the CRPD and MCA illustrate that rights can be incommensurable, much like the values they purport to reflect. Some kind of value judgement will inevitably be required to break the deadlock. A relational analysis of rights helps expose such competing values and could potentially help promote more transparent, democratic debate. Moreover, the realisation that rights do not always track conventional liberal categories and boundaries makes such an approach amenable to this type of critical evaluation of the types of relationships that are normatively desirable.

To return to the case study of Emma, a relational account of rights would likely offer a different analysis to the 'will and preferences' paradigm. The authenticity of her expressed wish to remain living with her father is something to probe and consider, particularly as the discourse of helplessness has clearly infiltrated her sense of herself and what is within her power. Though she can communicate her decision and reasons clearly, a relational analysis would question how her father's imposed restrictions (physical and social restrictions in terms of contact, psychological restrictions through narratives of disablement and extreme dependency) may compromise her autonomy and long-term agency. It would further probe the *assumptions* underlying the different responses. If one is appealing to Emma's right of autonomy, then a relational analysis would explore the

[93] MCA 2005 s.4(6)(a), (b).

normative justifiability of the type of relationship that is endorsed, either through intervening *or not*. It would also question whether the particular interpretation of that legal right has differential consequences on Emma. She is still owed deliberative respect regardless: indeed, the participation of Emma is central to any sort of engagement, whether Emma's wishes are ultimately implemented or overturned.[94] Ultimately, my point here is that a relational analysis would problematise the assumptions implicit in a 'will and preferences' response: it would examine how interpreting the right of autonomy as a negative, shield concept would preclude analysis of both the external and internal reasons behind one's choices, particularly as it is clear how legal recognition of the private relationship entrench her disablement, disempowerment, and inequality.[95]

Conclusion

In this chapter I have expressed scepticism that the 'will and preferences' interpretation of legal capacity can suitably replace the concept of mental capacity, mainly due to its problematic adherence to the private/public and individualistic assumptions in liberal rights. A relational analysis of legal rights, by contrast, provides a key interpretive prism through which a more relational concept of mental capacity becomes possible. This analysis helps expose ways that the law can structure relationships in normatively questionable ways. It likewise brings to the forefront much-needed debate about the underlying values that these rights seek to advance.

This is not to say that functional approaches to mental capacity as they currently stand fare much better than the external critique, as we see in the following chapters' examination of the MCA. The impetus of the common law tradition clearly pulls towards an individualist understanding, whilst the practical realities of impairment and disability necessarily pull in the opposite direction. Yet the argument for retaining the concept of mental capacity despite the 'will and preferences' interpretation goes beyond pragmatic reasons. The interdependent, relationally situated nature of individuals, the *intrinsic* challenges to particular impairments, are both acknowledged when the subject of rights is the embodied self. It seems to me dangerous and neglectful to adopt the abstracting, interpretive move

[94] Kong, 'The Space between Second-Personal Respect and Rational Care', pp. 433–67.

[95] Positive duties of support may well recommend safeguarding actions, such as her removal from her father in such circumstances.

that sweeps these realities about disability aside. As I discuss in subsequent chapters, the capacity to decide involves an interplay of unique, individual features and relational, social context, and recognition of this dynamic will be key to ethically informed, dialogical interactions that empower and enable individuals.

3

Relational Autonomy and the Promotion of Decisional Capacity

The previous chapter discussed how a relational analysis of rights draws attention to ways in which the law structures relationships in certain ways. The 'will and preferences' paradigm of the Convention for the Rights of Persons with Disabilities (CRPD), critically examined through the lens of feminist analyses of violence against women, illustrates the danger of adhering to individualistic and private/public boundary assumptions of liberal rights. Instead, a relational analysis of rights argues that individualistic legal interpretations in theory and practice can often mask abuse and social oppression. Later in the book (Chapter 5), I explore obligations and duties of support in these circumstances.

Confronting such problematic practices requires more substantive discussion about the values underlying liberal individual rights, particularly that of autonomy. The present chapter takes up this key challenge. Consider the following three scenarios:

1. Anne grew up in a strict, regimented household, with successful parents and older siblings. Throughout her childhood she experienced significant familial pressure to achieve academically as well as in ballet, the latter of which also discouraged weight gain. Since her adolescent years Anne has struggled with anorexia nervosa and has been in and out of hospital care. Now a young woman in her early twenties, she wishes to refuse treatment for her anorexia nervosa. Anne is fully aware of the consequences of her decision, but she cites how much the value of thinness matters to her.
2. Rob experienced his first severe schizophrenic episode when he was seventeen years old and was subsequently hospitalised. At the time his mother insisted that he remain within her care after his discharge and likewise resisted any social care assistance, claiming that she knew what was best for her son. Now in his fifties, Rob struggles to carry out basic tasks, such as changing batteries or adjusting the thermostat due to years of hearing narratives of his dependency and helplessness. Rob

resents the absence of basic practical skills, particularly as his elderly mother has become physically enfeebled. He feels a deep attachment to his mother and panics about the prospect of her dying. When offered a choice to moving to independent, supported living, he decides to remain in his mother's home, though he acknowledges that if had been left to his own devices without his mother's intervention, he likely would have accepted social care support for independent living much earlier in his life.

3. Joan has dementia and lives in a residential care home. Prior to her dementia, she had fulfilling relationships with her husband and children. She also had long-standing religious faith and expressed strong, consistent values about marital fidelity and the wrongness of premarital sex. Joan begins to get distressed when her family visits, and she increasingly isolates herself. Eventually, she refuses contact with her family and husband; at times she does not remember them, but when she does, she expresses that she has no desire to see them. Meanwhile she has also developed a close emotional attachment to, and has decided on beginning a sexual relationship with, another resident.

Amongst these three scenarios, whose decision expresses personal autonomy? Any or none of them could be our answer, depending on our perspective. As the cornerstone of liberal society, autonomy is defined as the ability to determine or govern oneself and one's life, independent of alien or external influences. How to flesh out this basic definition, however, is heavily contested. Individualistic accounts abound in mainstream bioethics, medico-juridical practice, and liberal philosophy, whereby autonomy remains a feature of the individual will and its criteria simply requires internal psychological consistency: that one's decisions accord with one's own preferences, values, and desires. From an opposing vantage point, this fixation on the individual neglects crucial relational, social factors that bear on the development and exercise of autonomy.

Medico-juridical adjudications of mental capacity reflect a similar confusion. We saw in the previous chapter how individual rights are often presumed to represent a protective sphere of negative liberty and the peremptory power of individuals. Even external critiques of mental capacity are not immune to these liberal assumptions. Latent confusion about autonomy contributes to conflicting notions of mental capacity in both theory and practice. On one hand, the link between internal cognitive features and the ability to decide is thought to determine whether one has mental capacity or not, or indeed, whether an individual retains his or her right of autonomy. Yet widespread recognition that some form of support is necessary for the realisation of decisional capacity amongst those with impairments militates against an individualistic conception of autonomy.

Building on the previous defence of a relational account of rights, this chapter focuses on an internal analysis of mental capacity that is grounded on a *relational* rather than individualistic concept of autonomy. It is absolutely crucial that we clarify what autonomy actually means given that mental capacity effectively operationalises this concept, functioning as the 'gatekeeping device' for individuals' right of autonomy. If autonomy is viewed individualistically – as an intrapersonal achievement of the individual – then it becomes much harder to justify capacity as a relational concept. Conversely, arguments for a relational account of autonomy will help establish the possibility of a relational account of capacity. What I am also looking for in my analysis here is a *shared, more inclusive account* of autonomy rather than the predominant tendency in philosophy to use examples of individuals with impairments as 'contrast' cases, mainly to illuminate what the autonomy ideal looks like for those without impairments.[1]

Thus, this chapter contends that autonomy requires intersubjective, relational conditions that develop and promote a range of different socially acquired skills. Autonomy conceived as such will equip us with vital tools necessary to meaningfully differentiate between the scenarios of Anne, Rob, and Joan, but likewise avoids dual-sided dangers of over- and under-inclusiveness in its criteria. I argue that two aspects are crucial to building up this relational model of autonomy: first, the phenomenology of *absorbed coping* brings to the forefront the perceptual skills involved in everyday interaction between body and environment, revealing the priority of the self's engaged intercorporeality over disengaged subjectivity. Second, absorbed coping supports a *procedural-relational* model of autonomy that can function as a plausible explanatory as well as normative backdrop to a more relational understanding of decisional capacity. Insights drawn from both phenomenological accounts of absorbed coping and the procedural-relational model of autonomy will make the concept of autonomy inclusive towards individuals with impairment, yet

[1] This is not to ignore that biological factors also bear on individuals' capacity to make decisions. However, I focus more on the social and relational aspects, mainly because current mental capacity legislation tends to already recognise (and at times, overemphasise) the causative nexus between biological causes (e.g. brain damage or mental disorder) and the inability to decide.

still contains demanding criteria so that autonomy continues to be an achievement.

Section I outlines the procedural-internalist approach to autonomy that remains entrenched in bioethical and medico-juridical contexts. Although this approach casts the net of autonomy wide enough to include most individuals, values, and goods, the normative minimalism of this approach is not only deceptive, but also contains descriptively false assumptions about personhood. Marina Oshana's substantive-relational approach to autonomy seeks to correct these problems, but Section II suggests that dangerous perfectionist and paternalistic implications result from the robust evaluative tools used to determine the sociocultural practices and values that are consonant or discordant with autonomy. Sections III and IV argue the problems afflicting these two models can be met through a phenomenologically informed, procedural-relational model of autonomy. The phenomenon of absorbed coping articulates the perceptual skills of everyday bodily engagement with one's environment. Our bodies, our ways of perceiving and thinking, are all socially inscribed, leading to the need to consider the relational turn in accounts of autonomy. But there needs to be a balance in this analysis: Diana Meyers' account of the socialisation process involved in authentic self-constitution and autonomy competencies successfully avoids the two extremes of valorising independence or self-reliance (at the expense of relational support) on one hand and untenable claims about the entirely social constitution of the self (which sanction perfectionist judgements about social or cultural values) on the other. Moreover, as we see in Section V, problems with overdemandingness (and thereby the exclusion of individuals with impairments) that tend to characterise other models are avoided through this scalar account of autonomy.

I. Autonomy in Bioethics and Liberal Theory

The value of autonomy has become pre-eminent in the medico-juridical context through the establishment of consent procedures in clinical practice and research. Lord Goff in *Airedale NHS Trust v Bland* describes how the right of autonomy has come to trump traditional values of medical paternalism and beneficence:

> [I]t is established that the principle of self-determination requires that respect must be given to the wishes of the patient, so that if an adult patient of sound mind refuses, however unreasonably, to consent to

> treatment or care by which his life would or might be prolonged, the doctors responsible for his care must give effect to his wishes, even though they do not consider it to be in his best interests to do so. . . . To this extent, the principle of the sanctity of human life must yield to the principle of self-determination. . . . The doctor's duty to act in the best interests of his patient must likewise be qualified.[2]

The implications of this perspectival shift are profound. Patient autonomy has now emerged as 'the most powerful principle in ethical decision-making in . . . medicine' and is 'the "default" principle' in regulation, bioethics, and theory.[3] Respect for patient autonomy requires specific conditions, however. As Lord Goff's qualification 'of sound mind' indicates, a certain threshold of psychological competence must be met for the right of autonomy to have priority. Influential bioethical theories of autonomy similarly require conditions of psychological competence. According to Beauchamp and Childress, autonomy as 'personal self-governance'[4] denotes the freedom to pursue self-chosen desires or plans independent from external constraint (such as coercion, control, and interference by others *or* by personal limitations). Such personal self-governance presupposes the possession of certain psychological and reasoning capacities, including 'understanding, intending, and voluntary decision making'.[5] Beauchamp writes, '[c]ompetence judgments function as a gatekeeping device for informed consent'.[6] To respect an individual's autonomy 'is to recognize with due appreciation that person's capacities and perspectives, including his or her right to hold certain views, to make certain choices, and to take certain actions based on personal values and beliefs.'[7] In other words, respect for autonomy demands deference to an individual's wishes

[2] *Airedale NHS Trust v Bland* [1993] 1 FLR 1026, 1035–1036 *per* Lord Goff.

[3] Paul Root Wolpe, 'The Triumph of Autonomy in American Bioethics: A Sociological View', in Raymond DeVries and Janardan Sbedi, eds., *Bioethics and Society: Constructing the Ethical Enterprise*, p. 43, quoted in Onora O'Neill, *Autonomy and Trust in Bioethics* (Cambridge: Cambridge University Press, 2002), p. 36.

[4] Beauchamp and Childress are emphatic that the principle of respect for autonomy *does not* override other moral principles and considerations, but only has prima facie standing (Tom L. Beauchamp and James F. Childress, *Principles of Biomedical Ethics*, 7th ed. (New York: Oxford University Press, 2012)). However, it is important to note the way that the principle of respect for autonomy has often overtaken other considerations in medico-juridical practice.

[5] Tom L. Beauchamp, 'The Four Principles Approach to Health Care Ethics', in *Standing on Principles* (New York: Oxford University Press, 2010), p. 37.

[6] Tom L. Beauchamp, 'Informed Consent: Its History and Meaning', in *ibid.*, pp. 72–3, also Beauchamp and Childress, *Principles of Biomedical Ethics*, p. 111.

[7] Beauchamp, 'The Four Principles Approach', p. 37.

so long as they meet a level of psychological competence associated with certain cognitive skills. Conversely, the wishes of those who fail to meet the competency threshold can be overridden on best interests grounds. The expression of and respect for autonomy revolves around the individual: its conditions require skills internal to the individual, whilst recognition of autonomy demands non-interference with another's subjective values, beliefs, and preferences.

Inspiration for this individualistic interpretation of autonomy could be drawn from traditional liberal theory. Mill famously defends a sphere of negative liberty against illegitimate outside interference and asserts, '[i]n the part which merely concerns himself, his independence, is, of right, absolute. Over himself, over his own body and mind, the individual is sovereign'.[8] This Millian formulation underlines civil and political rights fundamental to liberal society, such as the right to freedom of movement, thought, and expression. Similar to bioethical accounts, liberals make respect for autonomy – deference to an individual's choices, in other words – conditional on the presence of certain capacities and skills, fleshed out in terms of *competency* and *authenticity* criteria.

Competency conditions denote procedural mechanisms that ensure one's decision-making is internally and psychologically consistent.[9] They refer to the various capacities needed to deliberate and choose in accordance with one's values and include abilities of rational thought, self-understanding, and self-control, absent of coercive or pathological

[8] John Stuart Mill, *On Liberty* (London: Penguin, 1974), p. 69. MacDonald J cites Mill's point about the sovereignty of the individual in the 'sparkly case', *Kings College Hospital NHS Foundation Trust v C & V* [2015] EWCOP 80. Mill also qualifies there that the liberty principle 'appl[ies] only to human beings in the maturity of their faculties' (p. 69), a phrase that was explicitly appealed to by Mostyn J in *Rochdale Metropolitan Borough Council v KW & Ors* (Rev 1) [2014] EWCOP 45, para. 14, a striking judgment that directly challenged the Supreme Court's definition of liberty in *Cheshire West* and *MIG and MEG* cases (reported sub nom *P v Cheshire West and Chester Council and another; P and Q v Surrey County Council* [2014] UKSC 19, [2014] 1 AC 896). Of course there are other historical models of individualist autonomy, such as the Kantian approach. But Kant's account is too substantive and moralised to be a serious contender in the context of mental capacity adjudications. As we see in Chapter 5, however, his account of the Formula of Humanity can function as a fruitful justificatory source for duties of intervention in cases of neglect.

[9] Consistency in this context does not refer 'consistency over time', but consistency between the different hierarchical ordering between first- and second- (or higher-) order desires (explained further along in the chapter). This differs from those who would suggest that a standard of 'consistency over time' should be used as a measure of competency to decide. See Paul S. Appelbaum's response to Jeffrey Spike's letter, 'Patient's Competence to Consent to Treatment', *New England Journal of Medicine* 358:6 (2008): 644. Thanks to Gerben Meynen for bringing the reference to my attention.

influences. Authenticity conditions clarify what decisions can be attributed to the 'authentic self'. The notion of authenticity is considered the 'core to the liberal concept of capacity' in the medico-juridical context.[10] How to locate or define the authentic self remains deeply controversial, yet some form of self-knowledge and self-reflection is required to determine what is true and non-alien to the self. Authenticity conditions are met, therefore, when one chooses from desires, goods, or values that are one's own, validated through a reflective process of first-personal identification. If the desires, goods, or values that inform particularistic choice can withstand such reflective scrutiny, then the decision expresses the authentic, governing self.

The normative criteria of both competency and authenticity conditions are included in most liberal models of autonomy, though their internal specifications are fiercely contested along two spectrums. *The procedural-substantive* spectrum determines how much value content should be included in the criteria of autonomy, whilst the *internalist-externalist* spectrum concerns the extent to which features inside or outside of the individual (i.e. broader societal and relational factors) bear on autonomy (see Fig. 1). Put crudely, if autonomy is thought to revolve around the internal structure of the individual, the more competency conditions are stressed, whilst if external, relational forces are thought to affect autonomy, the specification of authenticity conditions will be increasingly important.[11] Yet proceduralist criteria need not assume autonomy depends on the internal structure of the individual – some accommodate the role of social forces and their impact on autonomy; substantive theories likewise are not always relational but can nonetheless focus on the interior content of an individual's motives.[12]

[10] Mary Donnelly, *Healthcare Decision-Making and the Law: Autonomy, Capacity and the Limits of Liberalism* (Cambridge: Cambridge University Press, 2010), p. 123. See also The Law Commission, *Mentally Incapacitated Adults and Decision-Making: Medical Treatment and Research* (Law Comm CO 129, 1993), p. 21, which recommended that 'a mentally disordered person should be considered unable to take the medical treatment decision in question if he or she . . . is unable because of mental disorder to make a true choice in relation to [the decision]'. The word 'true' was dropped in the MCA, though Donnelly argues that it is implied in the statute.

[11] See John Christman, 'Relational Autonomy, Liberal Individualism, and the Social Constitution of Selves', *Philosophical Studies* 117 (2004): 148.

[12] For example, Susan Wolf's influential substantive theory of autonomy suggests that autonomy amounts to normative competence – to act within and according to the reasonable constraints of right and wrong. See her book, *Freedom within Reason* (New York: Oxford University Press, 1990). Though Wolf's view is important, particularly in philosophical debates about autonomy, I do not consider her view in detail in the context of this chapter,

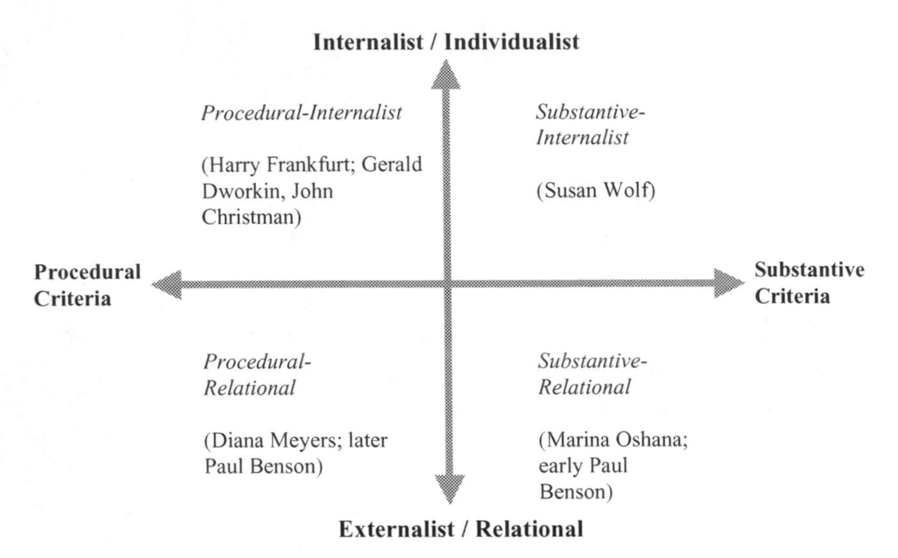

Figure 1. Two intersecting spectrums of autonomy.

Most models of liberal autonomy are hierarchical, in the sense that they examine the consistent organisation of its different components *within* the individual. According to the procedural-internalist account, our motivation comprises first-order desires and second-order volitions.[13] Competency and authenticity conditions are met when first-order desires and second-order volitions reflect internal consistency and procedural independence (where I identify choices and desires as my own through a process of reflection). For example, I might have a first-order desire to smoke cigarettes, but I have a contradictory, higher-order volition that values long-term health. My second-order volition cannot endorse my first-order desire to smoke: to do so would contradict my own self-consciously held values and commitments and therefore go against my authentic self. The fact that I cannot reflectively endorse my choice to smoke is what makes me non-autonomous, not that smoking cigarettes is a bad or objectionable desire in itself. In focusing on whether a person's second-order volitions – or a person's 'true self' – can identify with one's

particularly as it is unlikely this theory tracks the underlying presuppositions that are implicit in medico-juridical interpretations of capacity. The moralised inflection of her view makes it problematic to apply in such a context.

[13] Harry Frankfurt, 'Freedom of the Will and the Concept of a Person', *Journal of Philosophy* 68 (1971): 5–20. For another formulation, for a similar formulation, see Gerald Dworkin, *The Theory and Practice of Autonomy* (Cambridge: Cambridge University Press, 1988).

first-order desires, the normative criteria of autonomy remains neutral towards the actual content of one's desires, commitments and values; whether they are worthwhile is irrelevent. In Dworkin's words, '[w]hat is valuable about autonomy is that the commitments and promises a person makes be ones he views as his, as part of the person he wants to be, so that he defines himself via those commitments. But whether they be long-term or short, prima facie or absolute, permanent or temporary, is not what contributes to their value.'[14]

The inclusive or 'eclectic'[15] spirit of this model's normative minimalism has clear appeal: *more* rather than *fewer* individuals could qualify as autonomous if its normative criteria look at the procedural dimensions of critical endorsement rather than the substantive content of actual decisions. Different choices, values, and conceptions of the good life are warranted respect, regardless of their dubious value to others. This value-neutral stance contributes to the model's individualist focus in an important way. So long as we take ownership of our own volitions and desires, their originating source doesn't matter: whether our second-order volitions are the product of oppressive social norms or indoctrination does not immediately bear on our autonomy. Setting aside real issues about pathology and diagnosis, the decisions of Anne, Rob, and Joan from the preceding examples could all be considered autonomous according to this account. Suppose that Anne has assessed the risks and benefits of refusing treatment for her anorexia; her lower-order decision is thus consistent with her self-consciously held, higher-order values and goals. Her decision fulfills the criteria of procedural independence.[16] Rob's decision to remain with his mother might appear to fall short of the authenticity conditions initially, particularly in light of his regret of not doing otherwise. But suppose Rob has a second-order desire of personal security in addition to that of becoming more independent: according to the fairly thin normative criteria of the individualist model of autonomy, we could

[14] Dworkin, *Theory and Practice of Autonomy*, p. 26. See also Alexandre Erler and Tony Hope, 'Mental Disorder and the Concept of Authenticity', *Philosophy, Psychiatry, and Psychology* 21:3 (2014): 219–32.

[15] Christman's terms.

[16] Indeed, this conclusion can be a real problem in the context of refusal of treatment for eating disorders. Camillia Kong, 'Beyond the Balancing Scales: The Importance of Prejudice and Dialogue in *A Local Authority v E and Others*', *Child and Family Law Quarterly* 26:2 (2014): 216–36, as well as Jacinta Tan et al., 'Competence to Refuse Treatment in Anorexia Nervosa', *International Journal of Law and Psychiatry* 26 (2003): 697–707, note how individuals with anorexia nervosa often display high levels of procedural independence and internal consistency between first-order desires and second-order volitions.

interpret Rob's lower-order decision as consistent with these overarching values, of which have simply won out over the value he places on having a more independent life. This conclusion is perfectly consistent with the normative demands of this model because no substantive criteria evaluate the comparative authenticity of competing second-order desires. And finally, we may very well attribute the sudden radical shift in Joan's desires to changing second-order values, where marital fidelity and close family ties may have diminishing importance in light of her new existence within a residential care home.

The procedural-internalist model's seeming inclusivity helps explain its widespread appeal; however, it appears vulnerable to dual problems of overdemandingness and oversimplification. Overdemandingness in this context refers to the unrealistic standards individuals must meet in order to qualify as autonomous.[17] Standards of critical endorsement and consistency idealise an overly demanding picture of wholehearted, integrated selfhood[18] that comprises introspectible, unified states, absent of apathy or ambivalence about our choices. Such integration and deep commitment is unattainable for many, whether or not one is affected by mental impairment or cognitive disturbance. Selfhood and practical agency are far more complex. Our epistemic resources – and by default, our ability to know ourselves fully – are limited. Divided, unconscious motives; neurotic, addictive habits; choices with only superficial rather than wholehearted commitment – all complicate the process between decision and action.[19] Ignoring these complicated aspects of our agency restricts the spectrum of autonomous action, making autonomy an either-or scenario. The procedural-internalist model's inclusivity would appear to be jeopardised if it does indeed assume this overdemanding picture of the undivided, all-knowing self.

Oversimplification refers to the glossing over of important complexities and deviations from standard cases that then render the account irrelevant and unmeaningful. The problem of oversimplification in this case would be even more serious. The procedural-internalist model appears to lack the requisite evaluative tools to determine meaningfully when its own normative criteria have been met, given that the authenticity of one's

[17] Rather than the standard association the term has in criticisms of consequentialist ethics.

[18] Harry Frankfurt, 'Identification and Wholeheartedness' in *The Importance of What We Care About* (Cambridge: Cambridge University Press, 1998), pp. 159–76.

[19] See Erler and Hope, 'Mental Disorder', p. 225ff, for a sceptical view of whether hierarchical ordering of desires is possible when one has an attitude of ambivalence.

higher-order volitions cannot be ascertained. At what point does the process of critical endorsement stop?[20] The threat of an infinite regress is ultimately symptomatic of the model's crude ahistorical and asocial character. As a result, individuals who choose a life of subservience, slavery, or abuse could be deemed autonomous. Regress can be avoided only when we consider historical and socialising forces, pernicious or otherwise, in the competency and authenticity conditions. Issues of personal history and socialisation become unavoidable once authenticity conditions seek to probe *how* our values, desires, or volitions have been acquired.

But are procedural-internalists really vulnerable to these charges? They neither have to claim such deep levels of critical endorsement nor necessarily ignore the importance of personal history and socialisation on autonomy. Indeed, a historical account of procedural-internalism might sidestep both criticisms. John Christman acknowledges explicitly that different processes motivate individuals and their thought processes; at times manipulation and constraint are self-inflicted or the result of one's implicit approval. This means that '[w]e often are not so self-aware, nor do we always proceed in rational, self-controlling ways'.[21] He argues that autonomy must therefore 'leave some room for the unreflective and the automatic in life', so long as these unreflective dimensions do not stem from repressive, heteronomous states.[22] This leads Christman to suggest that autonomy requires only *surface-* rather than *deep-level*, historical understanding of one's higher-order values, utilising the test of *non-alienation*. This test makes the authenticity condition not simply an issue of positive endorsement or identification, but of both *negative* cognitive judgement as well as affective reaction, in terms of being 'constrained by the trait and wanting decidedly to repudiate' in light of our autobiographical, historical narrative of self.[23] In other words, Christman's historical account, on one hand, accepts the fact that reflection on higher-order values can often evade us, partly because such values themselves function as the orientating lens through which we come to make judgements in the first place; they form the prior structure which grounds our situational choices. Should the test of authenticity rely on us adopting some kind of

[20] Gary Watson, 'Free Agency', *The Journal of Philosophy* 72 (1975): 205–20; Irving Thalberg, 'Hierarchical Analyses of Unfree Action', *Canadian Journal of Philosophy* 8 (1978): 211–26; Marilyn Friedman, 'Autonomy and the Split-Level Self', *Southern Journal of Philosophy* 24 (1986): 19–35.

[21] John Christman, *The Politics of Persons; Individual Autonomy and Socio-historical Selves* (Cambridge: Cambridge University Press, 2009), p. 139.

[22] Ibid., p. 140. [23] Ibid., p. 144.

god's-eye, fully introspectible view, it would be unrealistic indeed. Instead, what Christman calls 'minimally adequate'[24] reflective judgement draws upon a historical sense of ourselves. Such judgement comprises both cognitive and affective components and expresses autonomy through consistency – when it is 'repeated over a variety of circumstances' and 'instantiate[s] the characteristic mode of thinking' within 'a variety of settings and contexts, always yielding neither alienation nor rejection.'[25]

Proceduralist-internalists like Christman can therefore reject a time-slice view of endorsement, shifting the focus of autonomy away from a particular moment of identification and towards the conditions of, and consistent expression during, the process of desire-formation.[26] In this way, a historical account might likewise sidestep worries about oversimplification. Christman acknowledges that it is impossible to scrutinise the authenticity of one's volitional commitments without understanding their underlying history for the person. We can test authenticity through three conditions: first is a counterfactual, hypothetical test of non-alienation to probe one's resistance to the *process* of developing a desire, value, or commitment and determine whether one did, or would have, objected to the conditions that led to their adoption. The second condition requires that the development of these values and desires did not or would not have occurred under conditions where self-reflection is obstructed. The final stipulation is that such self-reflection is rational, with no self-deception.[27] The ability to provide some kind of historical genealogy for our volitions not only halts the infinite regress of criticial endorsement, but also helps determine the authenticity of higher volitional sources, where sources of endorsement that are distorted through brainwashing, mind control, or manipulation, are distinguished from those that originate in the authentic self.

Similarly, Dworkin challenges existential notions of selfhood and acknowledges that social conditioning is an explanatory, causal condition of autonomy. He states, 'the notion of decision or choice is implausible as a description of how we acquire our motivational structures. We simply find ourselves motivated in certain ways and the notion of choosing, from ground zero, makes no sense. Sooner or later we find ourselves,

[24] Ibid., p. 154. [25] Ibid., pp. 152, 153.

[26] See John Christman, 'Autonomy and Personal History', *Canadian Journal of Philosophy* 21 (1991): 1–24; 'Relational Autonomy, Liberal Individualism, and the Social Constitution of Selves', *Philosophical Studies* 117 (2004): 143–64.

[27] Christman, 'Autonomy and Personal History', p. 11.

as in Neurath's metaphor of the ship in mid-ocean being reconstructed while sailing, in mid-history.'[28] With this in mind, Dworkin argues for the necessity of procedural independence – namely the ability to take ownership for the way in which others have shaped one's motivational structure as well as reflective and critical faculties. External social influences can subvert as well as promote or improve second-order critical endorsements – the important point for procedural independence is if one can view these, as well as the type of person one is, as being one's own.[29]

These modifications of the procedural-internalist model initially appear to help approximate the relative autonomy of two of the preceding cases in a more refined manner. The importance of the genealogical context is obvious in the case of Rob: he may himself recognise how his mother's influence contributed to decisions he retrospectively disowns. This might well compromise his procedural independence, though we would need to assess whether this retrospective repudiation wins out on balance over other values he purports to hold. Moreover, Joan's repudiation of her family and previous values could very well be viewed as autonomous choices according to Christman's test of non-alienation.

But these assessments reveal that the charge of overdemandingness might still apply, particularly in light of the strong individualism at the core of historical versions of procedural-internalism. Identification remains an individualistic endeavour, concerned with how one makes sense of oneself, either through positive endorsement or its negative counterpart of non-alienation. On one hand, Christman is clear that the sustained critical reflection characteristic of non-alienation means to accommodate the fact of relatively unconscious influences and challenge views that only deep reflection will do. Nevertheless, its overdemanding character remains. According to Christman, the type of reflection characteristic of autonomy requires oneself to ask 'whether I can take this as part of my ongoing autobiography, looking at my past as well as projecting into the future, and avoid the feelings of resistance and rejection characteristic of alienation'.[30] Indeed, Christman clarifies that he is 'considering a person with a settled and fully functioning value framework' who possesses the reflective skills to guarantee 'agential authority' that ensure reflections 'echo our personalities, our characters, our embodied selves, and other settled (and central) modes of our identities'.[31] On that basis, our initial

[28] Gerald Dworkin, 'Autonomy and Behavior Control', *Hastings Center Report* 6 (1976): 25.
[29] Dworkin, 'The Concept of Autonomy', pp. 212–13.
[30] Christman, *Politics of Persons*, p. 154. [31] Ibid., pp. 148, 149.

presumption that Joan's sudden change of heart and values passes the test of non-alienation would be in doubt.

That autonomy requires *some sort of* mitigation of social influences remains a background presupposition: even as individuals are embedded within the social process, there is a lingering suspicion that these processes are somehow *external* to functional, autonomous expression. Aside from the obvious cases of repudiation or revulsion at certain instances of social or relational influence, their impact will be more incidental rather than constitutive of individual agency. Dworkin, for example, states, 'we always retain the possibility of stepping back and judging where we are and where we want to be.'[32] And though Christman might admit that the unconscious and social cannot be separated from the individual and the motivation for certain choices, his account of reflection remains overdemanding due to its emphasis on *subjective procedural* skills which uphold the epistemic and practical importance of the disengaged stance. We apparently don't exercise these abilities *through* other individuals. Yet for many with impairments, the opposite is the case, where engagement and dialogue with others can often have a simultaneously *constitutive* and *reflective* function. I discuss this in more detail in Chapter 6.

Suffice to say here, the consequences of *not* recognising this leads to rather unpalatable consequences. According to Christman, individuals who lack this ability of reflection are ruled out as 'heteronomous' and 'should not be counted in collective decisions... [they] might well be open to paternalistic care rather than egalitarian respect in our social dealings'.[33] This hints that the historical version of procedural-internalism fails to truly address the problem of overdemandingness: it might not require full self-transparency or unrealistic epistemic abilities, but it certainly struggles to accommodate the experience of individuals whose impairments impede reflection of the diachronic self – namely how our lives ought to have a narrative structure over time, how we must make sense of ourselves in the long-term – and rather problematically, sanctions paternalistic rather than respectful treatment of them in those cases.[34]

[32] Dworkin, 'Autonomy and Behavior Control', p. 25.

[33] Christman, *Politics of Persons*, p. 162.

[34] This is not to say that paternalism isn't warranted in any case. But as I suggest in my article 'The Space between Second-Personal Respect and Rational Care', assumptions about the threshold of competence can be problematic, as we see in examples like eating disorders.

Overreliance on subjective procedural skills in historical versions suggests that the issue of oversimplification hasn't been fully addressed either, where there is a failure to incorporate the process of socialisation *within* the normative conditions of autonomy.[35] Closer examination of Anne's case helps explain this point. Even with refinements of the historical account, Anne's refusal of treament appears perfectly consistent with her second-order value of thinness at all costs. Whether this decision is autonomous becomes more questionable with closer examination of the process by which she acquired her higher-order desire – *regardless* of whether she endorses or not past familial pressure to achieve academic success or the value of a particular body weight for ballet. The procedural focus still does not get to the heart of why this case is so troubling. Strongly egosyntonic disorders, like eating and personality disorders, usually reflect highly consistent values and desires, which likewise cohere with historical narratives of oneself.[36] The historical version of procedural-internalism fails to track how the *content* of these narratives and values may alert us to problematic relational, social, cultural influences in the very *formation* and *procedural expression* of one's authentic self-identity.

Christman goes in the right direction in weakening the condition of self-transparency to the requirement of 'minimally adequate' reflection on one's values, but at the same time, due to the subjectivist emphasis, makes it impossible to critically evaluate those situations where individuals experience no alienating reactions towards insidious relational circumstances. In other words, Christman's retreat back to the subjectivist core of autonomy means that we have no evaluative tools at our disposal to judge

[35] In discussing Dworkin and Christman together, I am not suggesting that their views are necessarily equivalent. Christman's criticisms directly challenge Dworkin's continued commitment to the condition of identification for procedural independence. However, these criticisms do not overturn epistemic or subjectivist assumptions regarding self-transparency as a condition of autonomy, which do not accommodate easily the unconscious process of socialisation and the ways in which it can impact, positively or negatively, the achievement of autonomy.

[36] Egosyntonic refers to when an individual's specific thoughts, feelings, impulses, and behaviours cohere with, or are deemed acceptable, by one's self-image. Conversely, egodystonic denotes when aspects of one's thoughts, feelings, impulses, or behaviours are inconsistent with, or judged unacceptable, by one's self-image. Individuals with these types of egosyntonic disorders can still express ambivalent or contrary desires to their disorders. See Tony Hope et al., 'Agency, ambivalence and authenticity: the many ways in which anorexia nervosa can affect autonomy', *International Journal of Law in Context* 9:1 (2013): 20–36; also Enrica Marzola et al., 'A qualitative investigation into anorexia nervosa: The inner perspective', *Cogent Psychology* 2:1 (2015).

a person's historical and consistent identification with abusive relational or social contexts. Christman rightly acknowledges the complex constraints within certain states, particularly to do with relationships, where revising our choices would result in pain, effort, or great psychological cost, such as feelings of being in love or or devotion to one's children.[37] But he then uses these examples to show why 'slave' cases – those scenarios where individuals voluntarily enslave themselves – are no different, particularly if these choices reflect the consistent endorsement of values held by the diachronic self. He states, '[t]he fact that the values embodied in those reflective self-narratives are disturbing or self-abnegating does not gainsay the judgment that they are truly the person's own'.[38] But this presents a real problem: Christman goes too far in his subjective eclecticism, where the form rather than content of one's autobiographical narratives is what matters. If we hive off the content, we are deprived of the very evaluative tools needed to assess the social processes which contribute to a person's procedural skills of reflection. There are certain socially constituted narratives that fundamentally undermine individuals' procedural skills, regardless of their subjective stamp of endorsement – and, as the case of Anne shows, it seems to me that deeply self-abnegating narratives might just be an example of one.[39] Part of the reason the causal link between socialisation and the authentic self cannot be so neatly delineated is due to its impact on the development of personal identity in such multifarious ways. It may be so well integrated with our identity that where the 'authentic self' begins and socialisation ends is unclear. Social influences are not just causally implicated in who we are but, in many ways, form the basis of our procedural abilities to reflect and understand ourselves.

In sum, the procedural-internalist focus on the subjective, 'inner citadel'[40] of autonomy tends to disregard facts about human vulnerability and the relational support needed for developing and exercising self-determination, thus precluding an analysis of the complex interplay between independence, vulnerability, and relational connectedness

[37] Christman, *Politics of Persons*, p. 160. [38] Ibid., pp. 160–1.

[39] The self-abnegating character of certain narratives is key to distinguishing such cases from the person who takes up dangerous activities as a hobby (e.g. extreme sports). Individuals could come to serious harm in their pursuit of these activities in the same way that Anne's activities around food. However, people who engage in extreme sports will often have self-affirming rather than self-abnegating narratives which inform their hobbies (i.e. confidence and belief in oneself in the face of physical challenges). Thanks to Tony Hope for raising this example. I discuss this further below and in Chapter 6.

[40] Marina A. L. Oshana, 'Personal Autonomy and Society', *Journal of Social Philosophy* 29 (1998): 86.

within autonomy. There are two possible responses to these shortcomings: one is to stress the importance of externalist criteria in the formation of autonomy and focus attention on social, cultural, and political environment amenable to autonomy. As we will see from my analysis in the next section, this approach ultimately replicates certain problems within the procedural-internalist account. The other is to critically assess the socially constituted content of individuals' self-narratives and how these can promote or discourage their autonomy. Ultimately, this approach, combined with careful attentiveness to the phenomenology of absorbed coping, can best accommodate the embodied and relational dimensions of autonomy that are key for individuals with impairments.

II. Relational Autonomy – The Substantive Response

I have suggested thus far that the procedural-internalist model of autonomy is not appropriate as a foregrounding concept for a relational concept of mental capacity. Depending on whether the oversimplistic or overdemanding interpretation of its criteria is emphasised, this account can leave individuals with impairments and disabilities susceptible to either non-existent safeguards in obvious cases of relational neglect or abuse, or unwarranted, disrespectful paternalistic treatment. Dependence and impairment are not necessarily indicative of incapacity, nor do they signal that an individual's right to autonomy has been waived. The ethical obligations that follow from current legal paradigms, such as the Mental Capacity Act's (MCA's) discussion of decisional support in the Code of Practice, or the CRPD's endorsement of supportive decision-making, are hardly straightforward, nor are their accounts of this relationality fully coherent (as we have seen thus far). But even in their inchoate forms, these legal developments indicate that autonomy must assume a much more relational rather than individualistic temper in order to accommodate the experience of impairment.

Feminist theories of autonomy have echoed the need for this relational turn. A nurturing *interpersonal* environment can do a great deal to ameliorate vulnerabilities and empower the *intrapersonal* conditions of autonomy. Equally, socialisation and relationships can intensify these vulnerabilities and impede the development of autonomy. Neglect of these interpersonal conditions can ultimately harm the intrapersonal constituents of autonomy: repressing one's need for support and guidance can compromise the procedural skills that are necessary for authentic self-determination. Like the legal theorists examined in the previous chapter, feminist concerns about patriarchy and its internalisation amongst women have motivated closer scrutiny of ways that sociocultural conditions foster or limit capacities for autonomous choice. Two divergent approaches have emerged as a result: *substantive-relational* approaches attend to the actual content and values that make up the conditions necessary for autonomous deliberation and action, meaning that specific social practices are *constitutive* of individual autonomy. *Procedural-relational* approaches, by contrast, focus on how intersubjective, social conditions shape and contribute causally to procedural competency and autonomous identity-formation.

Consider Anne's scenario again: suppose that she critically endorses the familial and sociocultural values that stipulate women should be thin. Indeed, during her years in ballet (and prior to her clinical diagnosis of anorexia), her thinness conferred upon her some form of social status. Imagine that Anne has reflected on how her particular social conditioning might curtail some of her personal choices, but she nonetheless adopts these values as her own and takes personal responsibility for how they bear on her life choices. The procedural-internalist might conclude Anne's decision to refuse treatment is autonomous, but such an assessment would be counterintuitive from a substantive-relational perspective. From this standpoint, theoretical focus on the procedural conditions of choice fails to yield sufficiently robust evaluative resources to criticise the effects of oppressive socialisation on autonomy. For that, we need *more* rather than less content about the sorts of desires, values, and practices that are consonant or discordant with autonomy. Such content would form external, objective criteria that provide explicit answers to the question of *which* relational or sociocultural contexts enhance or detract from an individual's decisional autonomy.[41] So in the case of Anne, the fact that she subjectively endorses the value of thinness is irrelevant: the value itself is discordant with autonomy given that such a value emerged out of sufficiently oppressive social conditioning (i.e. her family, societal pressure on women to be thin, etc.), thus making her choices heteronomous.

Proponents of this approach provide different versions of the necessary content for a more substantive account of autonomy.[42] Marina

[41] Oshana, 'Personal Autonomy', p. 85.

[42] See Wolf, *Freedom within Reason*; also Paul Benson's early view of relational autonomy that endorses the condition of normative competence as a way of apprehending which norms or values ought to be endorsed: 'Freedom and Value', *Journal of Philosophy* 84 (1987): 465–86; 'Autonomy and Oppressive Socialization', *Social Theory and Practice* 17 (1991): 385–408.

Oshana provides a fairly robust approach to the effects of oppressive socialisation on autonomy. She argues that autonomy requires a range of internal and external conditions, including critical reflection, procedural independence, certain socio-relational practices and rights, and access to a full range of options.[43] The first two criteria, though similar to procedural-internalist accounts, can be directly affected or compromised by the last two conditions. For this reason, I want to focus on those. First, Oshana argues that procedural conditions of autonomy presuppose a relational, sociocultural context comprising certain substantive practices, values, and liberal rights. Such an environment 'must be free of whatever variety of factors destroy the psychological integrity of the person, and disable the person in her relations with others.'[44] Formal mechanisms cannot determine the psychological independence of persons nor distinguish between oppressive or non-oppressive social environments. Concrete social, political, and economic conditions must be met before procedural independence is even a possibility. The task of autonomy theory, then, is to articulate these conditions in more detail, according to which will then judge certain ways of life as inconsistent with autonomy. Second, the availability of a full range of options is a *constitutive necessity* for autonomy, meaning that an 'autonomous individual is one who is free to choose what sort of social relations to be in'.[45] More specifically, 'the option to choose nonsubservience must be available to the agent. Nor is an assortment adequate if the agent's choices are all dictated by duress (economic, emotional, etc.) or by bodily needs.'[46] Options dictated by necessity or restrictions of possible choice are incompatible with the normative demands of autonomy.

These two conditions are connected: if an individual's environment is composed of inherently restrictive social, political, and economic practices, this compromises the second condition that requires the freedom to choose the sort of social relations of which one wishes to be part. Oshana provides a striking example of a woman living under the Taliban, who freely chooses 'a life of utter dependence' in accordance with her religious values that, despite its restrictions on her dress,

[43] Oshana calls her approach 'social-relational'; however, in light of the taxonomy provided earlier, I will designate her approach as 'substantive-relational' for ease of comparison and to clarify her position on the procedural-substantive and internal-external spectrums of autonomy.

[44] Ibid., p. 94.

[45] Oshana, *Personal Autonomy in Society* (Aldershot: Ashgate, 2006), pp. 97, 106.

[46] Oshana, 'Personal Autonomy', p. 94.

movements, and practices, nonetheless 'provides her with a sense of worth, and satisfies her notion of well-being'.[47] Oshana argues that this woman cannot be called autonomous, as her

> genuine valuing of subservience or unquestioned adherence to religious tradition leads her to live a life of dependency. It diminishes the concept of autonomy to call such a human being autonomous in these conditions, for human beings are distinguished from other creatures precisely because of their deliberative and creative capacities. . . . This is not because *our* culture does not happen to value this way of life, but because it is a way of life that is inconsonant with autonomy, and autonomy itself is of considerable objective importance for all persons, whether or not it is of subjective importance to a particular individual.[48]

This example illustrates the claim that substantive conditions impact directly onto the procedural mechanisms of autonomy: procedural independence eludes the Taliban woman due to the absence of objectively necessary rights, liberties, and values. Her choices are already curtailed at the offset as a result. She is unable to make a genuine choice about the type of social relations she wishes to have because the option of nonsubservience is unavailable to her, and her endorsement of this restrictive existence therefore cannot be judged as autonomous.[49]

The advantage of this model of autonomy compared with more internalist accounts is the way it provides a concrete answer to questions surrounding *which* sociocultural practices and values obstruct the achievement of autonomy. This more detailed content might be especially welcome in the context of fraught capacity adjudications that turn on a care-recipient's relationships. Yet this model's practical applicability is called into serious question with closer scrutiny. Within the mechanisms designed to counteract oppressive socialisation is the same idealisation of

[47] Oshana, 'How Much Should We Value Autonomy?' *Social Philosophy and Policy* 20 (2003): 104.

[48] Ibid., pp. 104, 106–7. Susan Wolf might appear to make a similar argument in the context of responsibility, where she endorses a 'sane deep-self' view of moral culpability in 'Sanity and the Metaphysics of Responsibility', in Ferdinand David Schoeman, ed., *Responsibility, Character, and the Emotions: New Essays in Moral Psychology* (Cambridge: Cambridge University Press, 1987), pp. 46–62, a view that I have some sympathy with. Though Wolf's argument is similar to Oshana's in structure, to make a link between autonomy and responsibility requires an additional step. My view is that responsibility does not necessarily have to track the amount of autonomy one has; so even though we might promote the autonomy of individuals with particular impairments, we oftentimes nonetheless believe it is warranted not to hold them fully responsible. Thanks to Gerben Meynen drawing for drawing my attention to the similarities between the views.

[49] Oshana, 'How Much', p. 104. See Diana T. Meyers, 'Book Review of Oshana's *Personal Autonomy in Society*', *Hypatia* 23 (2008): 205.

subjective mastery and self-reliance embedded in procedural-internalist models of autonomy, thus implying that autonomy cannot coexist with dependency. In addition to sociopolitical rights and liberties, Oshana believes autonomy requires that an individual is not captive to 'psychological and physical disabilities, of the sort that prevent her from formulating and realizing her goals, and that she is able to *sustain herself without relying on the judgment and will of others*'.[50] Her more stringent interpretation of self-control denotes 'control over one's external circumstances'.[51] Yet the core essence of a relational model of autonomy is to overturn precisely such assumptions. Autonomy is meant to be embedded within, rather than oppose and transcend, facts of human vulnerability and dependency – which may very well preclude one's full mastery of or disengagement from external circumstances. This rings true especially for individuals with impairments. A learning impairment, for example, may fundamentally impede the ability to develop a healthy level of urban reserve needed to control and protect oneself from potentially harmful encounters with others. Or indeed, a person with borderline personality disorder may well epitomise disengagement when she goes into a dissociative state whilst engaging in self-harming behaviour. As Carolyn Ells eloquently puts it:

> With an experience of disability, a conception of self emerges that clearly and necessarily puts the self in relationship with its situation. While selves in relation to situation are not unique to lives with disability, those with disability are in a good position to remind us of this feature of all lives. For some of what goes unnoticed in typical unimpaired experience, such as embodiment and the extent of interconnectedness and interdependencies, emerges more distinctly in experience that includes disability. Moreover, it does so in such a way that reminds us that those connections and dependencies are there all along and for everyone.[52]

What Ells says here reinforces the need to think of autonomy in a much more nuanced sense. Indeed, the move to seek assistance and support through others can itself be an important indicator of autonomy skills – that we have a realistic notion of our constraints, limitations, as well as possibilities – compared with if we believed that we cannot or should not rely on others. Certain perceptual, cognitive, and deliberative skills eluding an individual in isolation might very well be achievable with reliance on others.

[50] Oshana, 'Personal Autonomy', p. 95, emphasis added. [51] Ibid., p. 82.

[52] Carolyn Ells, 'Lessons about Autonomy from the Experience of Disability', *Social Theory and Practice* 27 (2001): 609.

These statements are symptomatic of the even more worrying problem of paternalism. If procedural-internalist models of autonomy might be deemed too permissive and inclusive, Oshana's substantive-relational approach runs the opposite risk of being too paternalistic and exclusionary. Either result diminishes the full range and complexity of potential expressions of self-determination. Oshana explicitly endorses strong paternalism in order to restrict self-regarding harms that endanger personal autonomy.[53] Because there is 'little proof that adult individuals always know their own interests' and 'a person can be mistaken about whether or not he is autonomous, and about what his autonomy consists in' – as in the Taliban woman or contented but subservient housewife – 'restrictions on autonomously executed acts that eradicate one's dispositional or global autonomy can be upheld under a policy of judicious strong paternalism.'[54] But this begs the question: what justifies the epistemic superiority of certain people to determine when others are 'mistaken' about their own interests and self-understanding? Is it just that they incidentally endorse the same sociocultural values Oshana believes to be consonant with autonomy? The answer to this question revolves around the dubious claim that the capacity for autonomy amounts to a range of subjective epistemic, perceptual, and reasoning capacities which enable the *transcendence* of one's social context, without which one is unable to probe the authenticity of one's personal preferences. She states,

> Clearly, a number of individuals lack this capacity as I have described it. For example, a small child, an individual afflicted with Alzheimer's disease, and an insane person lack the rudimentary ability to be self-governing. Absent from all three is the characteristic of being a good "local sociologist," of apprehending the complexities of one's external environment, of consistently distinguishing malevolence from benevolence, and of comprehending the normative expectations of other persons and adapting one's behavior accordingly. Absent from all three is the power of self-appraisal and the ability to plan, to fix on preferences, and to function in a farseeing, deliberative, and self-protective manner. All three are creatures for whom certain forms of supervision and protection are appropriate.[55]

The sociocultural environment might constitute personal identity, but the attendant skills of a good sociologist seem to imply a measure of critical distance from one's social environment through perception and reason. Oshana states explicitly that critical reflection as a condition of autonomy demands one 'assumes the stance of a third-party in

[53] Oshana, 'How Much', p. 122. [54] Ibid., pp. 123, 124–5. [55] Ibid., p. 103.

appraising her motivations and actions, and the environment in which these develop'.[56] Elsewhere she stresses this process of disengagement further: '[i]f an agent is autonomous, she must have the latitude to distance herself – to step back – from socially given roles and practices, including those roles and practices that have, in part, made her. The autonomous individual must be free to question and judge these practices against her value and, if she decides, to opt out of them'.[57]

Oshana's account thus encounters a dilemma: either her assumption about the socially constituted self or the epistemic conditions of autonomy will have to be abandoned. The first assumption suggests that the social conditions make up the self in a strong sense. This is why the substantive content of social practices matter so much: on Oshana's view, they actually constitute the individual. Without the right practices, autonomy will elude the individual. But this strong thesis about the social constitution of the self rules out the possibility of transcending the social context through disengaged, critical reflection. From this direction, Oshana's conditions of critical disengagement and epistemic transparency have to give way. Conversely, if autonomy requires the sort of epistemic clarity and social disengagement that enables us to critically assess one's external environment, this presumes space between the authentic self and social conditioning, whether this be from the first-personal perspective of the individual herself or the third-personal perspective of the well-meaning paternalist who is enforcing autonomous practices. Oshana would then encounter the same problems of the proceduralist-internalist account surrounding questionable claims about objective critical reason and its ability to separate the authentic self from social practices. From that direction, Oshana's assumption about the socially constituted self becomes untenable.[58]

Oshana's highly perfectionist account of autonomy is understandable in some ways, given that she wishes to preserve a notion of autonomy as *achievement*. She sees as the task of any theory of autonomy to reveal how individuals in oppressive contexts lack personal autonomy, regardless of how much they might exercise it to some degree. Moreover, autonomy is thought to be a 'stable property' rather than a 'transient characteristic'.[59]

[56] Oshana, 'Personal Autonomy', p. 93. Also, Oshana, 'The Autonomy Bogeyman', *Journal of Value Inquiry* 35 (2001): 219.

[57] Oshana, 'The Misguided Marriage of Responsibility and Autonomy', *Journal of Ethics* 6 (2002): 261–80.

[58] John Christman similarly observes the individualist temper of Oshana's account in 'Relational Autonomy', p. 151.

[59] Oshana, 'Personal Autonomy', p. 95.

But at a more fundamental level, its paternalistic implications are highly unappealing. If strong paternalism can be used to intervene on the choices of a subservient wife, individuals with mental impairments would certainly warrant similar treatment. Yet theories of disability and practical legal developments are moving otherwise in their challenge to presumptions that mental impairment automatically means one lacks 'the rudimentary ability to self-governing'. In reality, the normative criteria of Oshana's model of autonomy would seem to exclude many of us, not just a small child, Alzheimer's patient, or those with mental impairment. Numerous aspects of our external environment are frequently beyond our epistemic investigation and subjective control; socialisation may run so deeply that attending to all of those features Oshana sees as constitutive of autonomy isn't possible without doing serious violence to the individual and her own self-understanding. As we see in the next section, the perceptual as well as social constituents of our identity often function as an unconscious backdrop to our sense of self and practical agency, thus eluding voluntary choice and analytical dissection, whether from a first-personal or third-personal perspective.

At this point, procedural-internalist and substantive-relational models of autonomy represent two possible extremes. At one end, we saw how the normative minimalism of the procedural-internalist model looks appealing but is in fact deceptive; its criteria are simultaneously over- and under-inclusive. Individuals who achieve psychological consistency, even within the context of deeply harmful egosyntonic mental disorders or oppressive social relationships, could be deemed autonomous. In both cases, our intuitions are likely (and ought to be) conflicted. This isn't to say that such individuals shouldn't necessarily have their choices respected; however, it seems somehow mistaken to suggest that these cases illustrate the ideal of autonomy. Yet those who often require the help and support of others to understand themselves and their choices (i.e. individuals with certain impairments) would be unlikely to qualify as autonomous, given the model's idealisation of independence and epistemic self-sufficiency. Equally, we need to be careful about veering too much in the opposite direction – where the environmental conditions are so demanding that an individual who endorses and chooses the values of a less than ideal social, relational context would be found to lack autonomy. At worst, both extremes impose an overly restrictive normative conception of autonomy and its attendant competencies, which then risks subjecting individuals with impairments to disrespectful paternalistic interventions. To avoid these pitfalls, the model of autonomy appropriate to mental capacity adjudications needs to incorporate three criteria: (i) competency conditions that are demanding enough to ensure functioning perceptual, emotional, and rational skills, yet do not automatically exclude individuals with cognitive impairments; (ii) an analysis of socialisation, culture, and relationships within an account of personal identity and self-determination whilst avoiding misleading claims that these factors constitute the self and first-personal perspective completely; and finally (iii) sufficient content to the procedural conditions of autonomy, which yet still tolerates and accepts diverse socio-cultural values and goods. This balance can be achieved through a twofold discussion of, first the phenomenology of *absorbed coping*, and second the *procedural-relational model of autonomy*.

III. Absorbed Coping

The skills of perception, affection, and bodily comportment are often viewed as the poorer second cousin to that of reflective thinking. A cognitive bias infiltrates even to the level of analysis of movement, where responses to anticipated objects or aspects of our environment is thought to be the result of some conscious willing or intention of the mind. Yet perception enables us to interact with our physical environment and has an utterly essential supportive function for cognition. What Samuel Todes refers to as 'poise'[60] or Hubert Dreyfus calls 'absorbed coping'[61] reflects how skilful bodily comportment is intrinsically geared towards everyday interaction and participation with the world around us. Coping refers to the ability to adapt, manage, and meet the demands of our environment or situation for one's benefit. But additionally, we can understand it in two different ways: one colloquial connotation denotes some kind of reflective, subjective disengagement from their environment: for example, someone 'copes' with the stress of an exam by consciously doing breathing exercises; I cope with an overly noisy, crowded train carriage by retreating into myself, playing music, or imagining I'm anywhere else but there. But the 'absorbed' adjective tries to push against the subjective, reflective disengagement implicit in these examples, so as to capture a second, deeper meaning of coping – namely how it demands *integration*

[60] Samuel Todes, *Body and World* (Cambridge, MA: MIT, 2001), p. 66.

[61] Hubert L. Dreyfus, *Skillful Coping; Essays on the Phenomenology of Everyday Perception and Action* (Oxford: Oxford University Press, 2014), p. 92 ff. Cf. Gerben Meynen's use of Michael Wheeler in 'Depression, Possibilities, and Competence: A Phenomenological Perspective', *Theoretical Medicine and Bioethics* 32:3 (2011): 181–91.

with the shared perceptual conditions of our environment. Our experience of trying to be at ease or comfortable ultimately doesn't depend on an *inward, detached* stance but instead relies on staying open, attuned, and intuitively reactive to the external world, through *pre-cognitive* perception and bodily movement.[62] Prior to any conceptual thought, our bodily interactions with the world allow us to gain some purchase on our environment.

According to Todes, constant orientations of our perceptual field confer stability to our bodily experience: our upright posture inevitably structures the scene before us in an up-down form; this vertical field functions as a necessary constraint in terms of what counts as skilful coping (i.e. meeting the balancing demands of this vertical axis).[63] Our body also has a front-back orientation that dictates what lies within our horizontal perceptual field. Our responses are geared towards what remains in the front rather than back of us.[64] Some things are seen closer or farther away, thus implicating a temporal field of perception – we can advance towards something that is further away, implying that '[t]he very structure of our forward-oriented active body always produces a future-field in which objects may appear ahead of us as still-to-be-encountered'.[65] The vertical and horizontal fields therefore determine the structure and constraints of our perception; they shape the anticipation of our pre-conceptual movement. Through these fields the world is opened up to us in particular ways, and we are able to situate ourselves within it.

Moreover, the habit-forming character of perception is a necessary feature that helps stabilise our experiences, minimising the sense of disruption we feel when experiencing the new, and shapes our anticipatory responses accordingly. As Todes argues, we engage in responsive action to achieve a certain level of comfort in our environment and meet our needs; as we accumulate more experience through time, we become more skilled in meeting our needs and accommodating ourselves to the world around us, to make the unfamiliar seem familiar, the uninhabitable, more habitable.[66] The expectation that our perceptions conform to past

[62] This move is extremely significant (and at root, follows Heidegger). The 'absorbed' adjective resists the disengagement implicit in the epistemic turn in modernity and conceptions of the self. In *Sources of the Self: The Making of Modern Identity* (Cambridge, MA: Harvard University Press, 1989) Charles Taylor shows how this mode of disengagement has profound consequences for how we think of our moral and political frameworks.

[63] Todes, *Body and World*, p. 126.

[64] Ibid., pp. 126, 171, Dreyfus, *Skillful Coping*, p. 122.

[65] Todes, *Body and World*, p. 171. [66] Ibid., p. 81.

experiences is a crucial coping mechanism in making our environment manageable to us.

Importantly, the phenomenon of poise or absorbed coping focuses on conditions of improvement – the sense of *appropriateness* or relief of tension – as opposed to conditions of success or achievement of purpose.[67] There is an important flow to movements that characterise absorbed coping. Consider, for example, the process of digging a hole in soil. I position the spade in the ground and the force of my foot feels right and appropriate. But I might fail to make a sufficient dent in the soil because of the unexpected presence of a tree root beneath. Or I might feel like the spade isn't heavy, or my grip steady enough, yet still make a deep enough well in the soil for my purposes. In this way, conditions of success and improvement can come apart. My body might be initiated by a willing or intention to dig a hole, but the actual movements themselves are engaged within a mode of absorbed coping which hinges on me instinctively adjusting my body and using the spade in response to my perceptions to achieve some kind of equilibrium, to adapt spontaneously and appropriately to my sense of tension (i.e. a tree root where I did not expect).

This perceptual dimension therefore draws attention to ways we move and respond, which is neither meaningless nor deterministic, cognitive nor psychological. It articulates how 'we are directly responsive to the other-than-rational demands of our situation [and] the way affordances solicit one to act.'[68] When we are in the active flow of things there is no separation, no cognitive disengagement, between one's movement and one's surroundings.[69] But through active movement what we do have is awareness of the body at three levels: (i) the skilfulness of our embodied movement vis-à-vis circumstantial objects, (ii) the coordination of our body, which leads to (iii) the 'felt unity' of our body.[70] The irony is that our cognitive unawareness of these essential perceptual skills means we demonstrate our skilful use of them.[71] But the distinction between one's movement and surroundings appears when one senses a disagreement

[67] Dreyfus, *Skillful Coping*, p. 150. [68] Ibid., p. 118.

[69] One objection might be that this is indistinguishable from determinism. I do not wish to engage too much in the compatibilism/incompatibilism debate, as this is ancillary to my main argument. However, I think it will become clearer later in this chapter that I do assume a compatibilist position and that the active *adaptability* of one's self-conception will be key to asserting human freedom, even as that self-constitution might be causally determined in the first instance.

[70] Todes, *Body and World*, p. 206. [71] Dreyfus, *Skillful Coping*, p. 95 ff.

between the attempt to do something and the failure to do it; this leads to a loss of poise and disorientation in one's circumstances, making experience less endurable and disrupting the 'felt unity' of the body.[72]

This brief account doesn't do justice to Todes' nor Dreyfus' rich phenomenological account. Nonetheless, their relevance to the context of impairment is beautifully captured in the first-hand narrative of a severely autistic adolescent boy, Naoki Higashida. In his book, *The Reason I Jump*, Naoki answers a series of questions about autism, drawing acutely the dangers of taking for granted the perceptual skills of absorbed coping. Naoki illustrates the priority of such perceptual skills to the conceptual in embodied coping, where the intrusion of cognitive disengagement can fundamentally disrupt the flow and felt unity of one's body: 'The reason I can't run well once I'm aware of needing to isn't to do with nerves. My problem is that as soon as I try to run fast, I start thinking about how I ought to be moving my arms and legs, and then my whole body freezes up.'[73] He also affirms the significant orientational function of the up-down, front-back perceptual fields, of how impairment can disrupt these skills, and more vitally, how one's body adapts in search of equilibrium and poise in light of such disruption. His explanation of why he jumps illustrates the significance of the up-down orientation in perceptual skill – even more so for Naoki, as it serves as a crucial coping strategy for affective distress and the experience of bodily disunity. He describes the agony of his feelings 'when I can't act, even though I really, badly want to. It's as if my whole body, except for my soul, feels as if it belongs to somebody else and I have zero control over it'.[74] By contrast,

> When I'm jumping, it's as if my feelings are going upwards to the sky. Really, my urge to be swallowed up by the sky is enough to make my heart quiver. When I'm jumping, I can feel my body parts really well, too – my bounding legs and my clapping hands – and that makes me feel so, so good. So that's one reason I jump, and recently I've noticed another reason. People with autism react to feelings of happiness and sadness. So when something happens that affects me emotionally, my body seizes up as if struck by lightening. 'Seizing up' doesn't mean that my muscles literally get stiff and immobile – rather, it means that I'm not free to move the way I want. So by jumping up and down, it's as if I'm shaking loose the ropes that are tying up my body.[75]

[72] Todes, *Body and World*, p. 70.

[73] Naoki Higashida, *The Reason I Jump*, trans. K. A. Yoshida and David Mitchell (London: Sceptre, 2007), pp. 117–18.

[74] Ibid., p. 68. [75] Ibid., p. 76–7.

Naoki further explains that kids with autism 'borrow' someone else's hand to make sense of the front-back orientation, because 'they can't tell how far they need to extend their arms to reach the object. They're not too sure how to actually grab the object either, because we have problems perceiving and gauging distances.'[76] But even as he struggles to gauge the distance of circumstantial objects, Naoki describes his constant wandering as both an ineffable interest in things emerging within his horizontal perceptual field and a persistent search for a place where he can become accustomed, that is habitable to him:

> I dash off as soon as I spot anything interesting. . . . We don't really know where we ought to be. . . . Simply put, people with autism never, ever feel at ease, wherever we are. Because of this, we wander off – or run away – in search of some location where we *do* feel at ease. While we're on this search, it doesn't occur to us to consider how or where we're going to end up. We get swallowed up by the illusion that unless we can find a place to belong, we are going to be all alone in the world. Then eventually we get lost, and have to be escorted back to the place we were at, or the person we were with before. But our uneasy, unsettled feeling doesn't go away.[77]

Two key implications emerge from this phenomenon of absorbed coping, evocatively captured in Naoki's narrative: first, this mode of perceptual engagement is *prior* to the disengaged mode implicated in the type of conceptual, cognitive reflection so highly valued in theories of autonomy. In emphasising the immediate interaction between perception and the surroundings, we move further away from a *meditational* to a *contact* picture of how we come to understand and engage with the world.[78] Since Descartes, this meditational picture holds powerful sway amongst philosophers: namely the view that our interactions with the world, and what we know about the world, are ultimately constructed through reflective and rational cognitive activity. But the phenomenology of absorbed coping reveals that the conceptual does not, in the first instance, mediate our epistemic and practical skill. Taking seriously how conceptual thinking is itself '"embedded" in everyday coping'[79] goes further than simple acknowledgement of its mere secondary role. Everyday coping itself forms the 'infrastructure' of our conceptual thinking; it has a fundamental, grounding quality which has characterised our lives from infancy to adulthood, whether one is impaired or not.

[76] Ibid., pp. 83–4. [77] Ibid., pp. 129–30.

[78] Hubert L. Dreyfus and Charles Taylor, *Retrieving Realism* (Cambridge, MA: Harvard University Press, 2015), p. 93.

[79] Ibid., p. 53.

Take, for example, the conceptual skills involved with the 'diachronic self' that so often forms a condition of autonomy. Without orientating perceptual skills that make us receptive to this temporal structure, the demands of diachronicity can be difficult. Naoki describes that '[t]ime is a continuous thing with no clear boundaries, which is why it's so confusing for people with autism. . . . You can't capture the passing of time on a piece of paper. The hands of a clock may show that some time has passed, but the fact we can't actually *feel* it makes us nervous'.[80] He goes further, '[w]e're anxious about what condition we'll be in at a future point, and what problems we'll trigger. People who have effortless control over themselves and their bodies never really experience this fear.'[81]

In other words, absorbed coping is normative and an achievement: it is an equilibrium we seek in our perceptual, embodied interactions with the world, and there is a sense of satisfaction when we manage to achieve it.[82] Attentiveness to the features of embodied poise helps balance out the cognitive bias underlying so many philosophical discussions of autonomy and its constitutive epistemic, practical skills. And offsetting this bias heightens our attunement to the extent in which individuals with impairments are engaged in this search for equilibrium and accommodation in their environment – where actions that might seem heteronomous or irrational at first glance, or violating conditions of diachronic narrativity, for example, may themselves be instances of skilful absorbed coping in light of their embodied reality. For instance, an outside observer might cite Naoki's wandering as evidence that his actions lack any reasoning; they seem like random actions with no intention behind them. Yet his narrative demonstrates very clearly how his wandering is an instance of his search for bodily poise; his wandering in itself expresses skilful coping in light of his perceptual reality, to accustom himself to his environment. I have more to say about this in Chapter 6, about how attunement to these perceptual constituents of experience enables one to better understand and interpret the actions and choices of those with impairments.

The contact rather than meditational picture implicit in embodied coping has a second, equally profound implication: namely it challenges the priority of the *subjective* domain by showing how an outward orientation beyond the internal confines of the subject is more naturally aligned with the earliest moments of absorbed coping. Dreyfus and Taylor write that '[t]he idea is deeply wrong that you can give a state description of

[80] Naoki, *The Reason I Jump*, p. 95. [81] Ibid. [82] Todes, *Body and World*, p. 264.

the agent without any reference to his or her world (or a description of the world qua world without saying a lot about the agent)'.[83] Contact and engagement with the world together lead to the 'interweaving of the bodily, the symbolic, and the narrative and propositional', which 'illustrates . . . the inseparability of the life and human meanings in the stream of human existence.'[84]

In our bodily comportment, social meanings are imprinted and expressed at the level of our movement within our environment and interaction with things.[85] Even more crucially, it represents our 'intercorporeality' or bodily attunement to others. Dreyfus and Taylor use the example of a baby's imitation of her mother's smile, prior to any observations about her own face and their resemblance. They continue that '[t]he sense of possible intentions in one's own body and the perception of these in others are primordially understood as belonging to the same type, and the likeness is lived in the communion of a two-person game, before it is grasped as a classification.'[86] This intercorporeality illustrates that even at our preconceptual stage we are oriented outwards, geared towards interaction, to 'enter into dialogical rhythms and common actions with others, playing endlessly respective games of hiding and revealing of concealment and surprise', which in turn, 'prepares the ground for the future postures of intimacy and distance, of common action and separate projects, which shape the geography of later social life.'[87] Much less than affirm the priority of the subjective, the internal domain, or a stance of disengagement, embodied coping reiterates the outward orientation of our perceptual skills, of how our body takes on a social significance. Our theoretical focus changes as a result. We move away from the subjective and internal, towards the interactive space between selves and their surrounding relationships and environment. As we will see below in the next section, this should reconfigure our concept of autonomy: although it may not entirely displace the 'subject'-focus of autonomy, it should nonetheless validate the relational turn in our analysis of the concept.

To summarise, this discussion about absorbed coping is crucial for two reasons: first, it is a key first step towards establishing a more shared conception of autonomy, balancing out the cognitivist bias that tend to exclude individuals with impairments. Instead, absorbed coping helps incorporate the bodily and perceptual experience of such individuals more fully. Second, absorbed coping is a necessary, though not

[83] Dreyfus and Taylor, *Retrieving Realism*, p. 94.
[84] Ibid., p. 117. [85] Ibid. [86] Ibid., p. 118. [87] Ibid.

sufficient, condition for autonomy: it works in concert with how we develop awareness about the social constituents of our self-conception, as I discuss below. The discussion thus far reveals how we have an intrinsically outward orientation to our bodies, making us receptive to our physical and social environment and helps shape the self-narratives that develop later in life.

IV. Procedural Relational Autonomy

The preceding discussion has explored the perceptual skills of absorbed coping and how it reveals the temporal priority of our affective and perceptual skills, intercorporeality, and engaged stance towards the world. These constituents of absorbed coping must be viewed as important contributions to autonomy, particularly if we are to make this account sensitive to the experiences of bodily impairment. They further help justify the relational as opposed to individualistic turn, where the 'self' or 'subject' in autonomy will still be important, but it must be grounded in the interactive nature of absorbed coping.

Diana Meyers forwards a model of relational autonomy that is sensitive to these phenomenological observations whilst achieving an optimal balance between the factors internal and external to the self. As Fig. 1 shows, the procedural-substantive distinction captures a spectrum of value-neutrality or value content, ranging from Dworkin and Frankfurt on one end to Oshana at the other. I refer to Meyers' theory as 'procedural-relational' for ease of comparison, though keeping in mind that Meyers herself (alongside later Benson) calls it a 'weak substantive' approach. This approach posits that some value content is necessary (though insufficient) for autonomy, whilst making space for individuals' varied commitments.[88] Meyers also structures autonomy around authenticity and competency conditions similar to procedural-internalists. '[T]he core of the concept of personal autonomy is the concept of an individual living in harmony with his or her authentic self',[89] she writes. Such harmony will presuppose the same aforementioned procedural abilities of critical self-inquiry and rational deliberation. However,

[88] Meyers, 'Book Review'. Cf. Benson's view in 'Feminist intuitions and the normative substance of autonomy', in J. S. Taylor, ed., *Personal Autonomy: New Essays on Personal Autonomy and Its Role in Contemporary Moral Philosophy* (Cambridge: Cambridge University Press, 2005), pp. 133–7.

[89] Diana T. Meyers, *Self, Society, and Personal Choice* (New York: Columbia University Press, 1989), pp. 49–50.

the first-personal, individual standpoint is only half of the picture of autonomy: its necessary counterpart is the social context that situates and influences, positively or negatively, the development and exercise of autonomy. How Meyers fleshes out this claim differently from substantive-relational models will become clear in a moment.

But first, her thesis that the achievement of procedural independence will depend on the nature of surrounding socialising influences marks an important departure from procedural-internalism. The latter recognises socialisation's causal role in forming personal identity, but implicitly assumes that autonomy revolves around the ability to disengage from these social influences through the process of critical reflection, endorsement, or tests of non-alienation. By contrast, Meyers argues that autonomy is developed and exercised *within* relationships, *alongside* the process of social conditioning. Authenticity and competency conditions are not just internal features of the free will, but a repertory of *socially acquired* practical, perceptual, cognitive, and emotional skills – what Meyers calls 'autonomy competencies'. These include the resolve to act in accordance with one's judgements, an 'ingrained disposition to consult the self, a capacity to discern the import of felt self-referential responses as well as independent beliefs, values, and goals, and a capacity to devise and carry out conduct congruent with the self'.[90] These competencies enable one to differentiate between acceptable or unacceptable situational constraints in one's choices as well as exercise 'a willingness coupled with an ability to deal with symptoms of discord within one's self.'[91] Success or failure to cultivate and exercise autonomy competencies frequently depends on environmental factors that are out of an individual's control in the first instance.

Meanwhile, the procedural criterion of critical endorsement demands both internal *and* external consistency. Skills of self-understanding and knowledge, sensitivity and attunement to our core values, are required, as is knowledge of how our situational preferences fit within the shape of our lives. However, the subjectivist turn of procedural-internalists is absent: for Meyers, 'self-knowledge should not be identified exclusively with introspective revelations'.[92] The social, cultural, and relational determinants of autonomy are more thoroughgoing and unconscious than procedural-internalists assume and, at times, the substantive-relational models of autonomy. The latter accounts are both preoccupied with the 'impossible' task of 'how people can become fully knowledgeable about

[90] Ibid., pp. 83–4. [91] Ibid., p. 55. [92] Ibid., p. 179.

themselves and their environment',[93] even if we assume deflationary epistemic skills necessary for surface-level reflection. For Meyers, the problem of autonomy is clarifying 'the difference between introspection and self-conception that are so misleading that they compromise autonomy, and introspection and self-conception that are sufficiently revealing to support autonomy.'[94] Instead of focusing on the existing internal structures of the authentic self (the main preoccupation of procedural-internalist model) or the content of sociocultural structures (the main preoccupation of a substantive-relational model), Meyers analyses the *process of constituting an authentic self*, where social forces continually interact with our self-conception in the development of autonomy competencies.[95]

This shift in emphasis reveals two things: first, the authentic self is a *dynamic*, unfixed entity. Aligned with the constant adjustments at the perceptual level of absorbed coping, it is flexible and revisable, responsive to emergent situational challenges and intrapsychical dissonances. The authentic self both recognises and takes personal responsibility for how group identities and social norms shape one's cognitive and motivational structure.[96] Second and more importantly, the exercise and achievement of autonomy is necessarily an intersubjective and dialogical process. Autonomy competencies are socially rather than individually acquired, grounded in the type of nurturing relations-to-self that comes with having a positive vision of our worth and value.[97] The intrinsic connection between autonomy competencies and nurturing relations-to-self (such as self-trust and self-esteem) implies that relationships often mediate our authentic self-understanding.[98] It builds on the intercorporeality and outward orientation we inevitably experience at the perceptual level

[93] Ibid. [94] Ibid.

[95] Diana T. Meyers, 'Intersectional Identity and the Authentic Self?: Opposites Attract!' in Catriona Mackenzie and Natalie Stoljar, eds., *Relational Autonomy; Feminist Perspectives on Autonomy, Agency, and the Social Self* (New York: Oxford University Press, 2000), p. 154.

[96] Ibid., p. 159. [97] I discuss these nurturing relations-to-self explicitly in Chapter 6.

[98] Paul Benson stresses particularly the condition of self-worth (or the subjective sense of 'worthiness to act') for autonomy. See 'Free Agency and Self-Worth', *Journal of Philosophy* 91 (1994): 650–68, and 'Feeling Crazy; Self-Worth and the Social Character of Responsibility', in MacKenzie and Stoljar, eds., *Relational Autonomy*, pp. 72–93. As I see this, a very basic notion of self-worth (as in the sense of one's value or esteem for oneself) is a necessary but not sufficient condition for autonomy, but Benson's discussion of the concept tends towards an overly substantive, moralised concept that still remains bound up with a notion of 'normative competence' (or moral justification/answerability to others), though these later papers reveal a more moderate position than previously held. What Benson's papers establish well, however, is how this sense of worthiness to act can be disrupted through oppressive and manipulative practices that become internalised by an agent.

throughout our lives. Constituting an authentic self will be a receptive, passive process over which we have no control in part, especially early in our lives. A child has no control over the core period of developing self-esteem, for example. She is born into her familial relationships, and her habituating, educative influences are largely unchosen. Whether her developing sense of self is nurtured or damaged is ultimately contingent on the relationships and sociocultural institutions she simply inherits. This receptive dimension of constituting the authentic self can also heighten certain inherent vulnerabilities in a dependent relationship. Individuals with mental impairments, for example, will often be vulnerable to their carers' messages of empowerment or disempowerment. Exposure to, and internalisation of, negative messages in our formative years give cause for us to distrust the absorbed coping of our bodies, our reflective judgement, and ourselves more generally; in turn, greater challenges are encountered in the process of constituting an authentic self. By contrast, positive messages of self-worth and value encourage the type of self-reading and trust in ourselves that is conducive to embodied poise and self-knowledge, thus enabling a more responsive, confident stance to be adopted towards environmental, social forces in relation to our personal identity.[99]

On the other hand, constituting an authentic self will also be an active, dialogical, and reciprocal process. Dialogue with others facilitates reflection of our personal thoughts, feelings, and values through the interpretive lens of others, so that 'our subjective reflections about ourselves take on an objective and practical reality and significance.'[100] But this dialogical process isn't only one-way: the use of language implies a constant reciprocal movement between social world and individual. As Dreyfus and Taylor describe, language has an orienting, habituating function and is the 'medium of expression of certain human meanings, defining a certain social world with its constitutive sense of what is important.' Conversely, 'it is . . . what integrates me into this social world, and makes those meanings manifest and real for me, as a child growing up.'[101] Dialogue can of course consolidate the pernicious influence of certain social or relational

[99] Cf. Benson, 'Free Agency and Self-Worth', pp. 650–68. Also, for a more Hegelian inflection on the role of self-respect and self-worth in developing autonomy, see Joel Anderson and Axel Honneth, 'Autonomy, Vulnerability, Recognition, and Justice', in John Christman and Joel Anderson, eds., *Autonomy and the Challenges to Liberalism* (Cambridge: Cambridge University Press, 2005), pp. 127–49.

[100] Kim Atkins, 'Autonomy and Autonomy Competencies: A Practical and Relational Approach', *Nursing Philosophy* 7 (2006): 212.

[101] Dreyus and Taylor, *Retrieving Realism*, p. 117.

contexts, just as much as it can be potentially empowering. However, empirical research indicates that these potential dangers can often be mitigated through group deliberation because cognitive abuses and questionable reasoning have a better chance of being noticed.[102] Meyers argues, 'this insight argues for a less privatized, more conversational approach' towards the 'enterprise of self-discovery', and '[i]nstead of pretending that self-knowledge depends exclusively on one's privileged access to a strictly private realm of mental states, people need to solicit and attend to others' impressions and suggestions.'[103]

In other words, self-knowledge that is conducive to autonomy relies on epistemic fallibility: we might have views about our embodiment, preferences, character, and values, but we might also ignore – consciously or otherwise – inconsistencies, contradictions, and ulterior motives. Dialogue can redirect our attention to areas of intrapsychical tension triggered by the social or relational context, thus opening up different possible avenues of self-understanding. It could spark a more determined endorsement of certain social norms so that discordant values or life plans within oneself need to be altered, or active resistance to those norms. Dialogue therefore has a multifaceted function in the process of self-constitution: it enables us to develop our sense of self in the first place; it has the potential to clarify our own subjective reflections, as well as correct our epistemic shortcomings. Conversely, the propensity to disengage from dialogue can be an important indicator of underdeveloped self-understanding and autonomy competencies. Closed off, rigid, or impotent self-definitions may be symptomatic of oppressive social conditioning's harmful impact on autonomy, similar to the way that inflexible, disengaged bodily responses may indicate struggles to cope or adapt at the perceptual level. Indeed, as Westlund contends, a monological approach to one's reflective capacities is indicative of limited autonomy, compared with an individual who possesses an outward, dialogical disposition to hold herself answerable for, and responsive to, the critical challenges to her internal endorsements within the interpersonal context.[104] Intersubjective, relational responses,

[102] Meyers, *Self, Society, and Personal Choice*, pp. 182–3. [103] Ibid.

[104] Andrea C. Westlund, 'Rethinking Relational Autonomy', *Hypatia* 24 (2009): 26–49. I do think that this can be taken a bit too far as a condition of autonomy, however. Westlund is correct to note how passive responses to interpersonal challenges can express a disconnection between an individual's practical reasoning and the justification of her commitments. However, I hesitate to say that this automatically expresses a lack of autonomy, particularly as Westlund does not provide a full account of the sociocultural environment that may, in fact, warrant an individual's passivity or non-responsiveness.

rather than a retreat to the subjective, are needed to counteract impediments to authentic self-constitution. Conversation with others encourages the self-knowledge that is necessary for autonomy, through which we 'test our theories and beliefs about ourselves, learn things about ourselves and are forced to confront things about ourselves'.[105]

V. Autonomy as a Scale

In sum, the procedural-relational model claims two conditions need to be in place *prior* to the achievement of authentic critical endorsement. First are the perceptual constituents of absorbed coping discussed earlier. Second is a range of relational, dialogical conditions that help develop autonomy competencies and authentic self-constitution, the content of which is the focus of Chapter 6. Autonomy emerges therefore as more of a spectrum than an either-or scenario; it can be achieved partially depending on whether these different conditions are in place and the range of autonomy skills and competencies within an individual's repertoire. This allows Meyers to sidestep the problem of overdemandingness. Let me explain by way of comparison.

Oshana's highly stringent approach cites instances of occurrent self-control as examples of heteronomy, effectively negating her scalar claims about autonomy. Her view is that theories of autonomy need to show precisely how occurrent control represents a lack of autonomy.[106] Inspiration for this view comes from Dworkin's original freedom versus autonomy distinction as resting on 'local' and 'global' concepts respectively. Freedom refers to executive control at a particular time and situation, whereas autonomy depends on the assessment of one's 'whole way of living one's life'.[107] Thus, 'freedom is neither necessary nor sufficient for autonomy'; they are 'different concepts' with varying scope.[108] To claim otherwise wrongly conflates separate logical categories. In short, localised self-control simply is not autonomy.

I agree that freedom is not the same as autonomy. As I suggested in the previous chapter, freedom as an opportunity concept is not equivalent to the exercise concept; but they are related, in so far as the very nature

[105] Atkins, 'Autonomy and Autonomy Competencies', p. 212.

[106] See Oshana, 'The Autonomy Bogeyman'; Oshana argues that localised responsibility and global autonomy are logically distinct concepts, partly because each presupposes a different underlying conception of rationality. See her paper, 'The Misguided Marriage', pp. 261–80. Also see Meyers, 'Book Review'.

[107] Dworkin, 'The Concept of Autonomy', p. 211. [108] Ibid.

of what counts as an 'constraint' will rest on how we understand our motivations and background values. But in making freedom as an opportunity concept irrelevant, the either-or view of autonomy then limits the (i) scope of autonomous agency and (ii) evaluative tools at our disposal for approximating the autonomy of others. These will have important consequences for mental capacity adjudications.

First, autonomy as an all-or-nothing achievement occludes a range of possible self-determining or 'episodically autonomous' actions that express the selective exercise of certain competencies in the absence of global autonomy. Not only would such a view be overly demanding – to the point that we would wonder how the ideal of autonomy can have any traction with the messy, complicated business of our lives – it is questionable whether the phenomenology of everyday coping properly supports this view. Absorbed coping often occurs at the pre-conceptual level, yet its importance to the everyday functioning of individuals is undeniable: an individual like Naoki might struggle to express global, reflective skills premised on diachronic narrativity, but it would be wrong to characterise as heteronomous his coping, orienting responses, such as the moment he jumps or instinctively reaches out to others to gauge perceptual depth. The scalar view is vital to ensure that autonomy includes as necessary (though perhaps not sufficient) the skills of everyday coping. This level of coping isn't important only for those with impairments. At some point or another, many of us compartmentalise portions of our lives and experiences, such as the individual who continues working in a job she despises and is contrary to what she envisages for her life. To conclude that she lacks autonomy ignores her everyday acts of functioning and self-determining choices, even as she stays within the job she categorically disowns.

By accommodating the importance of preconceptual, absorbed coping – or instinctive moments of freedom – we cast the net of autonomous agency more widely, thus mitigating the exclusionary thrust of overly demanding models. It seems to me that important nuances would be missed if we automatically assumed that psychological discord indicates a complete lack of autonomy. Individuals can express 'pockets of autonomy – particular actions – and threads of autonomy – policies addressing specific problems – in a person's life.'[109] Both episodic and programmatic autonomy require some form of self-consultation to carry out their decisions: for the former, perceptual adjustments or deliberation

[109] Meyers, *Self, Society, and Personal Choice*, p. 162.

may very well remain at the mere surface, in terms of 'How do I relieve this tension between what I need and my body's limitations? What adjustments can I make? What do I want *now* in this scenario or discrete area of my life? How can I implement this choice?' Clearly, these fall short of the deeper, more thorough questions characteristic of programmatic autonomy, where one probes 'What do I really care or value in my life as a whole?'[110] Programmatic autonomy thus examines our overarching goals and talents we deem worth developing and pursuing; it requires what Taylor calls 'strong evaluation' – applying the language of worth or value.[111] Meyers still stresses the importance of programmatic (or global) autonomy over episodic (or occurrent) self-direction but each alone is insufficient for autonomy. The accumulation of episodically autonomous acts in different spheres cannot substitute or culminate into a life plan that the authentic self reflectively endorses. Conversely, episodic, particular acts can nullify one's programmatic vision or plan. Minimally autonomous individuals have a certain disposition to consult themselves and can act in an instrumentally consistent manner to some degree, but they have poorly developed autonomy skills and competencies.[112] For example, these individuals might live a life of strong conformity (i.e. following gender, social, or cultural norms) and yet struggle to articulate whether these norms track something genuinely authentic to themselves. Medially autonomous agents achieve either episodic or programmatic autonomy, but not both (i.e. the fractured agent who compartmentalises areas of their life, or is deliberately erratic) and have a partial repertory of autonomy skills. Both the minimally and medially autonomous agents lack the full range of coordinated competencies that help unify episodic and programmatic control.[113] According to Meyers's procedural-relational picture, both the minimally and medially autonomous agents may be warranted respect for their discrete acts or choices because these often express some, if not all, autonomy competencies and independent skills.

Full autonomy aligns episodic and programmatic choices and expresses integrated but flexible self-constitution or self-narratives. But even then, we need to qualify this ideal of integration. On one hand, fully integrated, diachronic narrativity may very well elude some individuals with certain cognitive impairments. This could then sanction the disrespect of individuals' deliberation and choices through paternalistic treatment. On the other hand, the ideal of full integration can result in

[110] Ibid., p. 53. [111] Taylor, *Sources of the Self.*
[112] Meyers, *Self, Society, and Personal Choice*, p. 170. [113] Ibid.

counterintuitive assessments about the autonomy of individuals with egosyntonic disorders, as we saw in the discussion of Anne's case study earlier. If integration of episodic and programmatic choice is a necessary and sufficient condition for autonomy, then grounds for intervening in cases of serious self-harming but egosyntonic behaviours would be lacking. To track the complexity of such cases, we require evaluative tools that help us analyse how socialising forces can lead to disabling, self-abnegating narratives contrary to autonomy.

This latter problem highlights the second reason why Meyers's scalar approach to autonomy is so crucial. The evaluative tools at its disposal enable us to make sense of individuals' relative autonomy in these problematic cases where personal integration is evident but autonomy might be weak or lacking. There is no denying that we can believe and endorse egosyntonic but harmful conceptions of ourselves. Even if we have the potential to develop autonomy skills, internal and external factors interact and cultivate a self-conception that can either impede or promote our autonomy competencies. External obstacles include deception, peer pressure, disrespectful dialogue that perpetuates a narrative of individual worthlessness, or social structures that exclude certain values, beliefs, or characteristics that skew one's deliberation as a result.[114] This initially sounds similar to Oshana's claim that certain ways of life are categorically inconsistent with autonomy, especially if a full range of options is unavailable. But Oshana stresses certain substantive external conditions too much. Her reading of situations like the contented but subservient housewife or the Taliban woman misses the point in some respects. Certain legal rights and liberties are not baseline conditions for the possibility of personal autonomy. Their absence in oppressive circumstances can sometimes amplify expressions of autonomy through one's determined endorsement or dissent.[115] The converse is true as well: choices might appear deceptively autonomous due to their conformity with broader expectations about worthwhile values and goods in liberal society. Using the criteria of the substantive-relational model, it would be overly simplistic to conclude that more propitious external conditions of autonomy

[114] Ibid., p. 89.

[115] Meyers, 'Book Review', p. 205. Oshana in 'Personal Autonomy', p. 91, draws the conclusion that the conscientious objector is non-autonomous, for example. But this seems counterintuitive. Oshana tends to prioritise the external conditions of autonomy to the detriment of the internal, psychological conditions. I understand that these two types of conditions need to work in concert for autonomy, but it seems to me that one's active resistance to sociorelational conditions would suggest greater, not less, autonomy.

would amount to greater autonomy compared with the individual who strongly endorses or dissents from certain illiberal social practices.

By contrast, Meyers rightly emphasises, in her more procedural, interactive approach, how restrictive external structures impact more specifically on our internal autonomy competencies and self-constitution. These structures can skew 'chronic obliviousness to self-referential responses, awkwardness or rigidity in envisaging and appraising options, uncommunicativeness about one's needs, desires, values, and so forth, imperviousness to others' feedback, timidity about acting on the basis of one's own deliberations, and obstinate inflexibility in executing a chosen plan'.[116] Restrictive social structures can fail to cultivate in individuals perceptual and autonomy skills like personal flexibility, dialogical openness, epistemic humility, and the ability to address internal discord. Equally, internal aspects can thwart autonomy: mental disorder could distort our deliberation, or self-deception about our character or circumstances might be a more attractive option than engaging in the (sometimes) difficult process of constituting an authentic self. These internal impediments can also often originate in the interaction between external pressures and individual characteristics. They can manifest themselves in the failure to reach a decision that sits well with one's own self-understanding – or indeed, vice versa, where one's self-understanding and judgements have an air of certainty but fail to be enacted through lack of impetus, motivation, or will.[117] Guarding against these obstructions requires 'sensitivity to patterns of unfavourable self-referential responses and openness to the sceptical remarks of one's friends, on the one hand, coupled with a willingness to let one's imagination soar and to listen to others' inventive advice, on the other.'[118]

This invites a more nuanced assessment of the relative autonomy of each of the earlier case studies. If we return to the example of Anne, Oshana's account might question her autonomy regardless of the egosyntonic, coherent nature of her willing, but it tends to miss the mark as to why. The necessity of substantive values, such as freedom to choose otherwise or political rights, for example, do not provide plausible reasons for Anne's lack of autonomy. According to the procedural-relational approach, it isn't a foregone conclusion that Anne lacks autonomy entirely – her display of intrapersonal procedural skills cannot discount her from showing minimal or medial levels of autonomy, on one hand. On the other, how socialising forces inform the destructive, self-abnegating

[116] Ibid., p. 88. [117] Ibid., p. 89. [118] Ibid.

content of her self-conception – how they influence her highly rigid reasoning and everyday coping – indicates that her perceptual skills and autonomy competencies may fall short of full autonomy.[119] Likewise with Rob, we might look at how his conflicted views about personal independence might reflect society's broader expectations about worthwhile values and goods in liberal society. None of us is immune to the persistent assumption in liberalism that relational needs and goods are judged as less valuable compared with the goods achieved through independence and self-mastery.[120] So even if Rob were to choose to live elsewhere, away from his mother, in a bid for further independence, we might analyse whether this represents an authentic choice or mere compliance with mainstream societal values (with which we might or might not agree). Or we might look at how self-defeating narratives around him have developed a certain disabling way of understanding himself, leading to his underdeveloped ability to address discord within himself. If his decision reinforces internal conflict, it should be an intrapersonal indicator of the absence of full autonomy; yet he could remain medially autonomous nonetheless, in that Rob identifies how remaining with his mother helps him achieve some lesser values and goals. We nonetheless may have cause to question whether they express his authentic self, given his expressions of regret about his less independent existence. Indeed, part of constituting an authentic self is mature, more active reflection on the passive aspects of this process – to recognise and identify the unavoidable sociocultural influences that shape our development and to take responsibility for how they affect our future agency.[121] Either way of reading Rob's scenario would nonetheless indicate *weakened*, as opposed to the full absence of, autonomy.

[119] We might think that in Anne's case, the potential severity of her mental disorder would override these social considerations. The point of this analysis, however, is to highlight how socialising factors (once internalised) can lead to mental disorder that seems fully coherent with one's self-conception, making the genealogical history of one's self-conception particularly important. Thanks to Gerben Meynen for raising this objection.

[120] Anne Donchin, 'Understanding Autonomy Relationally: Toward a Reconfiguration of Bioethical Principles', *Journal of Medicine and Philosophy* 26 (2001): 375.

[121] Obviously this overlaps with Christman's historical approach. However, my interpretation of Meyers is that the process of reflection does not involve disengaging from the social context and environment, nor does it assume the ideal of self-transparency, given that unconscious forces and conditions continually influence the exercise of autonomy competencies. The important point is, rather, whether Rob has the dispositional propensity to consult himself and respond to felt dissonances between his preferences and values, and those of his relational environment.

The scalar approach of the procedural-relational approach moreover helps avoid the problem of perfectionism, typically encapsulated in the dominance principle. This principle posits that individuals ought to choose a course of action that guarantees more rather than fewer desired ends; it is therefore rational to prioritise our basic interests.[122] These include self-regarding ends such as well-being, survival, health, strength, intelligence, companionability, power, wealth, friendship, opportunities, and reputation.[123] The dominance principle further claims that certain goods amongst this list are necessary for autonomy; the ability to rationally weigh up basic interests with other competing values or interests is therefore part and parcel of autonomous decision-making. Certain goods (i.e. survival, health, measure of material goods) are presumed to have priority, as they are deemed necessary for sustaining autonomy. This is especially so if competing desires and values are either disconnected to the authentic self or reflect a skewed, deceptive self-understanding. The dominance principle is intuitively plausible but needs to be qualified. Meyers rightly argues that these basic interests are not always necessary for autonomy and their priority will depend on the particular context. Indeed, foregoing these basic interests in favour of other goods could be both autonomous and rational, especially when the opposing good has a firm connection to the authentic self.[124]

Joan's scenario is a case in point. Suppose that the good of healthy family life is a basic interest. From one standpoint, Joan's apparent sacrifice of a basic interest in favour of what appears a more recent, sudden, short-term interest is irrational: common sense would lend support for the dominance principle in this case. Whether the opposing values are autonomously held would be questionable. But if her decision has to do with her changing self-conception, where she is in the process of altering her understanding of herself in light of her new altered circumstances and environment, her choice to discontinue contact with her family and embark on a new sexual relationship would seem to reflect values and desires that can legitimately compete with basic interests. In that case, we would be hard-pressed to judge Joan's decision as nonautonomous without making more substantive, perfectionist judgements against her actual values.[125] Anne's case affirms Meyers's point even more acutely. From a

[122] Meyers, *Self, Society, and Personal Choice*, p. 102.

[123] Ibid., pp. 98–9. [124] Ibid., p. 105.

[125] One might argue that we ought to appeal to Joan's diagnosis to suggest her decision is non-autonomous. Whilst some argue that the wishes of the pre-dementia individual takes

perspective of the dominance principle, her refusal of treatment expresses the fundamental irrationality of rejecting basic interests (of life and health), and would further indicate a lack of autonomy. This argument is made even more compelling in light of the harmful body image and social messaging that typically underlie such treatment refusals. But again, the dominance principle risks being too perfectionist about the values at stake: cases similar to Anne's can recommend deference to one's subjective choice to forego such basic interests, particularly if to do otherwise is to inflict substantial psychological violence on the individual.[126] The latter is a particular danger in the context of egosyntonic disorders and impairments, hence why it is so important to adopt a scalar, non-perfectionist approach to assessing the relative autonomy in such cases.

The discussion thus far shows that adherence to the dominance principle is not an absolute requirement for autonomy. Yet connecting autonomy with the development of individual potentialities might give a different inflection to the dominance principle: goods that help guarantee individual self-interest should be chosen because this enables an individual's potential to be properly developed. Rawls's Aristotelian principle, for example, is meant to be an additional factor to be included in deliberation, stating that people 'enjoy the exercise of their realized capacities (their innate or trained abilities), and this enjoyment increases the more the capacity is realized, or the greater its complexity'.[127] This Aristotelian interpretation of the dominance principle assumes that individuals who actively seek to realise their potential can be deemed more authentic and true to their self-interest compared with those who do not. The achievement or underachievement of one's potential indicates therefore whether one is more or less autonomous. This line of reasoning presumes that potentialities and their realisation are self-chosen desires or values.

Ultimately, the Aristotelian version of the dominance principle is grounded in a worrying perfectionism that conflates autonomy with the

priority over the wishes of the post-dementia individual – see Ronald Dworkin, *Life's Dominion: An Argument about Abortion, Euthanasia, and Individual Freedom* (New York: Alfred A. Knopf, 1993) – there have nonetheless been persuasive arguments against this presumption. See Rebecca Dresser, 'Dworkin on Dementia: Elegant Theory, Questionable Policy', *Hastings Center Report* 25:6 (1995): 32–8 and Agnieszka Jaworska, 'Respecting the Margins of Agency: Alzheimer's Patients and the Capacity to Value', *Philosophy and Public Affairs* 28:2 (1999): 105–38.

[126] See *A Local Authority v E and Others* [2012] EWHC COP, [2012] COPLR 441, and my analysis in Kong, 'Beyond the Balancing Scales'.

[127] John Rawls, *A Theory of Justice*, revised ed. (Oxford: Oxford University Press, 1999), p. 374.

Protestant work ethic. Admittedly, autonomous individuals might well tend to develop their potentialities. But support for this claim falls short of a principle or requirement of autonomy: because autonomy requires a level of self-understanding and self-knowledge, attention will be drawn to these potentialities in the first place, which may or *may not* lead to a commitment to their cultivation and development.[128] Individuals do not necessarily choose their potentialities. If these potentialities do not cohere with an individual's self-conception, why would their neglect then be deemed less authentic? I might have the potential to be a brilliant mathematician due to some natural abilities with logic and numbers, but if it's a talent that I do not endorse or value or that isn't highly prioritised in my relationships or community, to allow these latent abilities remain underdeveloped, or developed only to the level that is necessary, could hardly mean that my authenticity is compromised. Individuals who might lack the desire to activate and develop their potentialities could still be acting consistently with autonomously held views about their self-interest. There are potentialities that appear like talents, gifts, or aptitudes (as in Rawls's Aristotelian principle), on one hand, and there are those potentialities that are essential to the development of autonomy competencies in the first place. We must avoid conflating the two if we are to conceptualise autonomy in a sufficiently broad manner to include a range of potentialities and choices.

Both versions of the dominance principle – the presumption in favour of basic interests and the Aristotelian principle – *presuppose* the abilities of self-understanding and critical reflection; such autonomy competencies are in fact *prior to* the deliberative application of the dominance principle.[129] The dominance principle itself is therefore not coextensive with autonomous decision-making, but a deliberative mechanism that helps (i) prioritise basic interests aimed at enhancing one's autonomy for as long as possible, particularly when an individual's opposing values and desires are non-autonomous; and (ii) clarify an individual's desires and values in a particular situation, whether it be the conscious cultivation or neglect of one's aptitudes, talents, and gifts. But clearly, the dominance principle as a deliberative mechanism falls short of a criterion of autonomy that sanctions substantive perfectionist judgements against an individual's chosen goods and values.

The normative appeal of the procedural-relational approach to autonomy should be apparent in three ways. First, the concept of autonomy

[128] Meyers, *Self, Society, and Personal Choice*, p. 116. [129] Ibid.

competencies, defined as a range of socially acquired perceptual, emotional, and reasoning skills needed in constituting an authentic self, means the normative criteria of autonomy is not so demanding as to automatically exclude individuals with psychological impairments. Meyers rightly acknowledges that autonomy is more a matter of degree than an either-or scenario, and even the achievement of episodic self-direction will involve the cultivation of both skills of absorbed coping and autonomy competencies, through interpersonal support and encouragement.[130] Though Oshana too is sympathetic to the idea of autonomy as a matter of degree, her substantive approach remains focused on highlighting the *lack of autonomy* in certain conditions, even if there are superficial, occurrent expressions of it. Oshana could argue that if episodic control is possible without programmatic autonomy, this would suggest that we would need to respect the decisions of individuals who desire and choose slavery, drug addiction, imprisonment, or spousal subservience.[131] Or to use an example that is all too familiar to the mental health context, it would mean calling autonomous the care-recipient who chooses to remain within an overtly coercive, abusive, and therefore disabling relational context. Safeguards against these types of practices – even if an individual endorses them – would be insufficiently robust, hence why the right of autonomy should be restricted to those whose full-blown programmatic autonomy has been realised. But the paternalistic implications of Oshana's approach are equally, if not more, problematic. The procedural-relational approach of Meyers might imply weaker safeguards against certain practices discordant with autonomy, but it is on the whole more inclusive and attentive to the complexities of bodily impairment, recognising that decisional capacity can be context-specific.[132] Autonomy competencies and skills include a range of abilities that can be enacted in different spheres, with varying levels of reflections about one's identity and life plan. Oshana's worries might be justifiable, but the prioritisation of programmatic autonomy over episodic instances of coping and self-control would imply that many individuals who are incapable of reflecting on their diachronic selves would not necessarily deserve respect for their desires, choices, and values in a discrete context and domain.

[130] Ibid., p. 162. [131] Oshana, 'Autonomy Bogeyman', p. 215.

[132] As I argue in Chapter 5, Meyers's approach is consistent with Kantian justification for further safeguarding in oppressive and disabling relational contexts that amplify intrinsic vulnerabilities of individuals with mental impairments.

Second, Meyers makes realistic assumptions about the role of socialisation in the development of autonomy skills and authentic self-constitution, without making mistaken assumptions that personal identity is either reducible to or transcendent of the social context. The procedural-internalist model, on one hand, tends to overestimate the extent to which self-concepts can be voluntaristically taken up or abandoned. The substantive-relational model, on the other hand, is committed to contradictory premises about the social constitution of the self and the possibility of epistemic self-sufficiency. For the procedural-relational approach, complete self-transparency asks too much of individuals in the face of deeply entrenched intercorporeal and intersectional constituents of our identity. But this claim falls short of Oshana's essentialism about the entirely socially constituted self. Socialisation has a causal role in inculcating and facilitating qualities of flexibility, revision, and personal dynamism in light of changing circumstances. Attuned to the constant adaptive, orientational movements of everyday coping, autonomy at perceptual and reflective levels 'enables people to regard their self-concepts as sufficiently *provisional* to be able to *monitor and adjust* them as needed'.[133]

Finally, the procedural-relational model avoids the perfectionism implicit in substantive-relational accounts. The full possession of autonomy competencies need not mean we must actualise our full potential or that we are embedded within an ideal sociocultural context. Judging individuals as somehow ignorant of their own interests is 'to indulge in misplaced and paternalistic moralism'.[134] The body's role in our absorbed coping, our own epistemic fallibility, must be kept in mind in our judgements of others and their communities. Autonomy rests on specific qualities within the interaction between internal (subjective) and external (relational) determinants. Focusing our analytical attention on the development of autonomy competencies will raise key questions, such as whether the social and cultural context is such that first-personal revisions are permitted and encouraged on one hand, and, on the other, whether the individual has the confidence, deliberative ability, self-esteem (all of which involve perceptual, cognitive skills) necessary to initiate and stand behind these revisions. This basic evaluative criterion is consistent with a range of sociocultural values and goods, as well as individual choices that challenge the dominance principle. That capacity assessments are often triggered precisely in cases where a person deliberates in a manner inconsistent with either incarnation of the dominance

[133] Ibid., p. 187, emphasis added. [134] Ibid., p. 115.

principle makes this point particularly salient. In cases when an individual chooses against her basic interests or acts in a manner that squanders her potential, scrutiny should be focused on the presence or absence of autonomy competencies rather than her actual values.

Conclusion

Current mental capacity law seeks to extend the right of autonomy to individuals with impairments, and the discussion in this chapter has critically examined assumptions in mainstream procedural-internalist theories of autonomy. The phenomenology of embodiment and absorbed coping fly in the face of individualist assumptions, making its normative idealisation of self-reliance a gross misfit with the practical realities of how perceptual and autonomy skills are developed and exercised by those who live with impairments, as well as everyone else. Relational autonomy helps us better understand the possible environmental vulnerabilities that emerge as a result of these empirical realities. Absorbed coping highlights how our basic perceptual orientation implies intercorporeality, engagement rather than disengagement with our surroundings. Our bodies bear inscriptions from our interaction with the social world. This is true also of the reflective skills so crucial to autonomy, where social conditioning and relationships can enable or disable our abilities to make authentic choices. Equally, we need to be wary of presupposing an overly demanding and substantive account of autonomy in capacity adjudications – where its normative criteria automatically excludes a range of either sociocultural values and goods, or potential expressions of self-determination, which may just fall short of full autonomy.

As the symbolic gatekeeper of autonomy, the foregoing discussion should impact on our concept of capacity accordingly. The outcome of capacity adjudications implies autonomy is an either-or scenario. But the decision- and context-specific definition of capacity lends support for a more scalar view of autonomy, with varied intersections between episodic and programmatic means of control. For this task, the procedural-relational model of autonomy is deeply attractive: the possession of autonomy competencies does not necessarily entail the realisation of our full potential, and respect for our self-determination may be necessary even if our abilities have not been fully actualised. Autonomy remains an achievement without being overly exclusive or demanding, through which a person's autonomy competencies can be assessed through important interpersonal and intrapersonal indicators. Self-formation is entrenched within the social context, at both the perceptual level of everyday coping and the cognitive level of rational reflection. Oppressive sociorelational contexts will often justify closer scrutiny of an individual's choices. But the procedural focus helps guard against perfectionist and paternalistic impulses that, more often than not, still tend to determine capacity adjudications.

Relational autonomy as a backdrop to capacity adjudication will bear on the model of rationality that is presupposed in the functional test of mental capacity. The practice of capacity assessment tends to assume that individual reasoning can hive off environmental, relational, and social factors. Intrapersonal factors can and do bear on our ability to reason, but it cannot be the whole story, as we will see in the next chapter.

4

Procedural Reasoning and the Social Space of Reasons in Capacity Assessments

The previous two chapters have made a twofold argument, helping us develop different facets of a relational picture of mental capacity: so far, I have argued that mental capacity must draw upon a relational concept of legal rights and autonomy. Feminist analyses of violence against women expose the problems associated with individualistic rights interpretations and help resist the external challenge to mental capacity posed by the 'will and preferences' interpretation. Phenomenological accounts of absorbed coping, combined with the insights of the procedural-relational model of autonomy, make the concept of autonomy sufficiently inclusive towards individuals with impairment, yet demanding enough to make autonomy an achievement nonetheless. It also scrutinises the social process of self-constitution, drawing our analytical focus to ways in which socialisation can inscribe enabling or disabling narratives in our bodily existence and self-understanding. In short, Chapter 3 stressed the impact relationships have on developing the first-personal characteristics of mental capacity and autonomy.

In this and the next chapter, I want to extend this relational lens to the actual practice of mental capacity assessment and explore the more second-personal and agent-neutral aspects to mental capacity: namely what kind of reasoning is expected of individuals? On what grounds can we intervene in privately chosen but abusive, disabling relationships? How do we ensure that we take seriously the relationality of rights, autonomy, and mental capacity, without veering too closely towards the direction of paternalism? These kinds of concerns I explore here and in the following chapter.

This chapter takes the relationality that grounds our conception of rights and autonomy and extends it to reasoning and rational norms. I have argued thus far that the development, promotion, and exercise of autonomy occur through particular types of relationships, mainly because autonomy's constituent perceptual, affective, and reflective skills are socially acquired. These competencies include our ability to reason and deliberate – to make judgements about ourselves and the world around us, to make practical decisions about our lives. By the same token, then, our rational decision-making abilities will have a relational, social dimension. We don't reason within a vacuum, nor is it purely an introspective exercise – our deliberation is intrinsically responsive to our social environment. And as I argue in this chapter, the social, relational context determines which reasons are deemed valid or invalid.

This claim appears to fly in the face of how current interpretations of mental capacity understand rationality. Consider the following case: in 2013 a 37-year-old patient wished to terminate her pregnancy in the twenty-third week of its term. She had a history of bipolar disorder – of which the symptoms could be periodically controlled by medication; other times the relapse of her illness led to her compulsory detainment. During the first stage of her pregnancy she had acted in a manner that demonstrated 'every sign of wanting to keep this baby and of desiring to be a loving and caring mother to the baby.'[1] Perhaps out of this desire the patient stopped taking her prescribed medication for her disorder. At the same time, the mother and husband of the patient reported a complete reversal in her attitude towards the pregnancy, where she was now adamant about wishes to have an abortion. The patient had also become compulsorily detained under section 2 of the Mental Health Act 1983, during which her wish to have an abortion was repeated. Both psychiatric experts and her family argued the patient lacked capacity to decide whether to terminate her pregnancy on grounds that her paranoid thoughts prevented her from understanding information relevant to her decision (that her husband and mother would be supportive of both her and the baby), meaning she could not use and weigh information both in relation to the question of family support as well as the foreseeable consequences of the abortion.[2]

Surprisingly, Holman J reached the opposite conclusion, stating, 'even if aspects of the decision making are influenced by paranoid thoughts in relation to her husband and her mother, she is nevertheless able to describe, and genuinely holds, a range of rational reasons for her decision.'[3] These reasons included her unhappiness with her detainment and the prospect of giving birth in such circumstances, her anxiety about her ability to raise a child in light of her recurring bipolar disorder, and finally, the absence of regret about her decision to terminate an earlier

[1] *Re SB (A Patient; Capacity to Consent to Termination)* [2013] EWHC 1417 (COP), para. 20.
[2] Ibid., para. 39. [3] Ibid., para. 44.

pregnancy. Although the judge's assessment might appear puzzling, it seemed to hinge on the patient's reasoning satisfying certain rational norms, such as consistency in claims and logically valid comparative thinking. Or more specifically, the capacity assessment was based on a *procedural model of rationality* (which focuses primarily on the process through which a decision is arrived at or the abstract structure of rationality), as opposed to a *substantive model of rationality* (which surrounds the qualitative value or goodness of the deliberative premises or outcome).

Arguments have been made that the appeal to procedural norms of rationality in capacity adjudications is both appropriate and necessary given the statutory aim of protecting the autonomy of individuals with impairments.[4] Yet the procedural model explicitly endorsed in medico-juridical tests of capacity is increasingly recognised as deficient. From the perspective of relational autonomy I defended in the previous chapter, the intrapersonal focus of the procedural model's normative framework neglects the perceptual, social, and dialogical contexts that help or impede a person's reasoning capacity and the exercise of autonomy competencies. From a practitioner's point of view, moreover, the individualistic, introspective focus of the procedural model of reasoning is fundamentally at odds with the unique relational, supportive conditions necessary to enable impaired individuals to exercise choice. The importance of support is repeatedly emphasised in both the Mental Capacity Act 2005 (MCA) itself and its Code of Practice. And rather worryingly, the minimal standards of procedural rationality can be met even in highly oppressive and abusive relational contexts. This model of reasoning is therefore of little help for those capacity adjudications having to distinguish when support melds into coercive force, making judgments of capacity based on a patient's relational environment appear arbitrary and unprincipled.

I want to approach this practical problem by challenging the procedural model on three fronts: the first two critically examine the theoretical presuppositions internal to its rational norms, whilst the third makes a more positive argument for going beyond its focus on the abstract structure of one's deliberation in capacity assessments. I argue that assessing the decisional support given to individuals with impairments demands a

[4] Appelbaum and Grisso, 'The MacArthur Treatment Competence Study I' and 'The MacArthur Treatment Competency Study II'; Jillian Craigie and Alicia Coram, 'Irrationality, Mental Capacities, and Neuroscience', in Nicole A. Vincent, ed., *Neuroscience and Legal Responsibility* (Oxford: Oxford University Press, 2013), pp. 85–109.

more substantive model of reasoning that the functional test of capacity currently rejects.

Section I outlines the use of the procedural model in medico-juridical competency tests. I then make the following challenges in Section II. First, the individualistic and introspective focus of the procedural model rests on a mistake about norms of justification. Robert Brandom's discussion of the social space of reasons shows that the norms of procedural rationality are *socially affirmed*: validity and justification hinges on the practical attitude and endorsement of others. Certain beliefs and claims are given a normative stamp of approval within this interpersonal space. Second, the social space of reasons operates in a circular manner, influencing the entire process of when beliefs and claims are conceived, to when they are explicitly articulated. Even setting aside the contentious issue of whether propositional logic and inference is in fact coextensive with rationality,[5] this capacity to make logical connections and comparisons can be (and often is) encouraged through intersubjective dialogue and relationships.

These first two claims are more explanatory in their challenge to the internal presuppositions of the procedural model. But recognising the intrinsically interpersonal character of the space of reasons makes it necessary to answer broader normative questions, such as the following: What reasons and commitments are justified in specific contexts or spaces of reasons? When different contexts collide (as they do when medico-juridical capacity assessors consider the quality of relationships surrounding individuals), why should some reasons and commitments have priority over others? Section III explores a version of proceduralism that might potentially accommodate Brandom's insights in an attempt to resolve these issues, but which ultimately fails because of its strict adherence to standards of content- and value-neutrality. As I discuss in Section IV, answers to these questions will involve examining the normative and practical value, rather than the mere form or structure, of the social space of reasons enframing an individual's self-understanding. For this reason, I make a case for what Charles Taylor calls the *articulatory-disclosive* mode of reasoning to be used in capacity adjudications, of which I define subsequently. Section V will apply this account of reasoning to two cases from the Court of Protection.

[5] This is called into question by Gilbert Harman in *Reasoning, Meaning and Mind* (Oxford: Oxford University Press, 1999), pp. 21–32. For criticisms of Harman's view, see José L. Bermúdez, 'Normativity and Rationality in Delusional Psychiatric Disorders', *Mind & Language* 16:5 (2001): pp. 467–8.

I. Competency Assessment and the Procedural Model of Reason

Most findings of incapacity hinge on the MCA's third pillar of the functional test: the ability to 'use and weigh' information. Yet amongst the criteria of capacity, it permits broadest interpretive latitude. More precision might be inferred through closer examination of clinical tests of competency.[6]

According to Appelbaum and Grisso's method, clinical tools assessing competency to consent to medical treatment have four components: (i) understanding, (ii) appreciation, (iii) reasoning, and (iv) expressing a choice. The first and last categories correspond directly to the MCA's first and fourth pillars of the functional test: understanding examines a patient's ability to comprehend and paraphrase information that is relevant to the patient's condition or disorder, the recommended as well as alternate treatment options, and their respective risks and benefits. Expressing a choice involves the ability to communicate a decision with the use of tools and support if necessary. In the clinical context, appreciation refers to the insight to relate relevant factual information to one's own situation as well as the ability to attach certain values to the risks and benefits of treatment.[7]

The third component, reasoning denotes the thought process through which one manipulates information to reach a decision, specifically the capacity to engage in logically coherent comparative and consequential thinking. The form or structure of reasoning is assessed, not 'the accuracy or reasonableness of the premises employed in the reasoning'.[8] This pillar thus seeks to remain neutral towards the substantive or epistemic value of the actual beliefs, premises, or decision,[9] though the weighing process may be impeded by a variety of cognitive and emotional impairments such as psychotic thoughts, dementia, extreme phobia or panic, anxiety, euphoria, depression, and anger.[10]

There are arguments that the component of rationality has been entirely excluded from the law in England and Wales; however, upon closer examination, this perception is based on a confusion between a procedural and

[6] M. J. Gunn et al., 'Decision-Making Capacity,' *Medical Law Review* 7 (1999): 294.

[7] Appelbaum and Grisso, 'The MacArthur Treatment Competence Study I', p. 115, and Grisso et al., 'Study II', pp. 115, 128–9; Paul S. Appelbaum, 'Assessment of Patients' Capacities to Consent to Treatment', *The New England Journal of Medicine* 357:18 (2007): 1836.

[8] Grisso et al., 'Study II', p. 129. [9] Ibid.

[10] Appelbaum, 'Assessment of Patients' Capacities', p. 1836.

a substantive interpretation of reasoning. Emphasis on the procedure as opposed to the content of reasoning in Appelbaum and Grisso is echoed in an influential pre-MCA case, *Re MB*.[11] Although *Re MB* is typically upheld as an example of how the criterion of reasoning is *excluded* from legal tests of capacity, what the judgment actually says is that irrationality may be *indicative of*, rather than *coextensive with*, incompetence to consent. Irrationality denotes 'a decision that is so outrageous in its defiance of logic or of accepted moral standards that no sensible person who had applied his mind to the question to be decided it could have arrived at it'.[12] However, 'it might be otherwise if a decision is based on a misperception of reality (e.g. the blood is poisoned because it is red). Such a misperception will be more readily accepted to be a disorder of the mind.' Similar to Appelbaum and Grisso, this judgment suggests that certain beliefs, compulsions, or misperceptions may impede the process of using information and weighing it in the balance in order to arrive at a decision. Quoting Lord Cockburn C. J., 'One object may be so forced upon the attention of the invalid as to shut out all others that might require consideration'.[13] Should a compulsive disorder or phobia impede belief in relevant information, doubts would be raised that the weighing process to reach a decision has remained unimpaired.

The case *Re MB* is instructive because it is typically upheld as the precursor to the criteria adopted in the MCA. Closer examination of this judgment suggests that the clinical categories of 'appreciation' and 'reasoning' are in fact condensed into the single category of 'use and weigh'. To be able to use and weigh information seemingly denotes the ability to, firstly, assign values or weights to certain benefits and harms so that one can commit and endorse a belief, and, secondly, generate the logical conclusion from one's beliefs and desires to choose a course of action in light of all other alternatives.[14] Drawing upon the *Re MB* judgment, Holman J in *Re SB* argues precisely along these lines, stating

[11] *Re MB* [1997] EWCA Civ 3093. [12] Ibid., para. 30 (3). [13] Ibid., para. 30 (4.b).

[14] According to Matthew Hotopf, 'The Assessment of Mental Capacity', *Clinical Medicine* 5:6 (2005): 581: 'There is a wide range of interpretation of appreciation, particularly as one moves away from overt psychotic symptoms towards the influence of emotional disorders, affect, and even culturally held beliefs. Although it is true that appreciation may be subsumed under the third criterion of the Mental Capacity Act definition (use of and ability to weigh information), the definition appears less elastic than Grisso and Appelbaum's.' However, it is unclear what Hotopf means when he claims that the application of appreciation is 'less elastic' in the MCA compared with Grisso and Appelbaum. On the contrary, juridical interpretations of use and weigh appear much more inclusive and flexible than the clinical interpretation of appreciation.

that 'if one does not "believe" a particular piece of information then one does not, in truth, "comprehend" or "understand" it, nor can it be said that one is able to "use" or "weigh" it. In other words, the specific requirement of belief is subsumed in the more general requirements of understanding and of the ability to use and weigh information'.[15]

These statements challenge arguments that the law in England and Wales has excluded the reasoning criterion in the test of capacity.[16] In reality, the judicial approach in both *Re MB* and *Re SB* reject explicitly an outcome-based, substantive model of rationality where the standard is whether the *content* of one's beliefs, reasons, or outcomes reflect particular normative values or goods.[17] Instead, the functional test upholds a *procedural* model of rationality as part of what it means to 'use and weigh' information. Reason, 'defined in terms of a certain style, method, or procedure of thought,'[18] is constituted by probabilistic and consequential modes of thinking which are guided by conditions of (i) internalism and (ii) logical consistency.[19]

Internalism makes the standard of rationality contingent on individual agents. Bernard Williams defines valid reasons as *internal* reasons: in other words, reasons are relative to an agent's present *subjective motivational set*, comprising an individual's subjective preferences, desires, or beliefs, as well as 'dispositions of evaluation, patterns of emotional reaction, personal loyalties, and various projects, as they may abstractly be called, embodying commitments'.[20] For example, think of the dictum, 'you ought to give to charity'. Maybe I have a personal commitment or desire to work towards ending poverty; or perhaps this is part of my religious creed. This gives me motivating reasons to give to charity accordingly. But if I didn't have any of such personal commitments, such a dictum will not inspire

[15] *Re SB*, para. 81.

[16] Gunn et al., 'Decision-Making Capacity', p. 295. Though Gunn et al.'s research pre-dates the MCA, they observe how the Law Commission's eschewal of rationality as a criterion of capacity accords with a contestable interpretation of *Re MB* [1997]. Even after the MCA's instantiation, they make the still valid point that interpretations of capacity rely on some sort of notion of rationality and reasoning, even as the common law has tried to set aside such standards.

[17] Craigie and Coram, 'Irrationality', pp. 85–6.

[18] Ibid., p. 86. There are numerous versions of procedural rationality. For my purpose I want to provide the most general and commonly agreed upon criteria to apply to the medico-juridical context. I discuss refinements to this model later in the chapter.

[19] Harman, *Reasoning*, pp. 18–20, also Bermúdez, 'Normativity and Rationality', pp. 465–8.

[20] Bernard Williams, 'Internal and External Reason', in *Moral Luck; Philosophical Papers 1973–1980* (Cambridge: Cambridge University Press, 1981), pp. 101–13.

me to suddenly give to charity. Internalism stipulates that action-guiding (or practical) reasons have motivational and normative grip on an agent only so long as they have traction with one's subjective motivational set. In short, subjective desires, preferences, and commitments initiate the deliberative process. Practical deliberation then amounts to instrumental rationality: roughly a means-end structure in which individuals determine how to satisfy their existing preferences and desires or fulfil their present commitments, attach relative weights and values to the various alternatives available for their satisfaction, and make a decision accordingly. This makes the content of deliberation indeterminate, flexible, and open-ended: the reasoning process can insert new actions to internal reasons and vice versa, where reasons can be added for given actions; similarly deliberation about the means required to satisfy an aspect of one's preferences could lead to its abandonment or the creation of new desires.[21] But regardless of how our actions or desires change, the key point is that we are ultimately responsive and receptive to reasons that are *our own* reasons, reasons that track something in our own personal commitments.

A second condition of the procedural model demands logical consistency: namely adherence to rules of deductive and inductive logic to generate valid inferences as well as eliminate manifestly contradictory premises. This includes the ability to draw comparative, transitive inferences so that preferences can be ordered and compared (A to B, B to C, then A to C). At the level of practical deliberation, the norm of consistency applies to one's practical reasons, reasons understood as comprising beliefs and subjective motivational sets or pro-attitudes. A syllogistic structure should guide one's instrumental reasoning, where according to Davidson, we imagine 'that the agent's beliefs and desires provide him with the premises of an argument.'[22] For example, the patent inconsistency of Kate who has the apparently strong desire to perform well in an exam, yet chooses not to study despite her belief that studying will enable her to perform well, may help pinpoint where her mean-end reasoning has failed.[23] Not only will the inconsistency be evident from a first-personal perspective, third-party *explanations* of Kate's

[21] Ibid., pp. 104–5.

[22] Donald Davidson, 'Intending', in *Essays on Actions and Events* (Oxford: Oxford University Press), pp. 85–6.

[23] Of course, the inconsistency could be down to deficits in an individual's motivational structure (such as *akrasia* or *enkrasia*).

actions will take the form of a logically invalid inference.[24] Moreover, the inconsistency would suggest that the desire to perform well must not be truly an internal reason – not part of her subjective motivational set – given that there exists no psychological link between the belief 'it is good for Kate to study' and what she is actually motivated to do.

The significance of these two procedural conditions is that it tries to make rational norms *intrinsic* rather than *extrinsic* to an agent's existing reasons and desires; it eschews the claim that the normativity of reasons stem from sources *external to* a subjective source, or 'by whether the agent is motivated to do what she (independently) ought'.[25] In other words, 'the independent variable must be some procedure of practical reasoning whose correct exercise can be specified internally and not by its approximation to some external, independently specifiable result'.[26] Procedural reasoning generates its own normativity rather than presupposes it. This means that it can *remain neutral towards content and value* even as it outlines the abstract structure of rationality. When we judge the rationality of individuals according to this model, we aren't evaluating whether her beliefs, desires, and commitments are right, good or worthwhile, but simply whether they pass the muster of procedural norms.

The finding of mental capacity in *Re SB* thus seems more understandable in light of this model of procedural rationality, whereby Holman J's interpretation of 'use and weigh' evaluates the patient's process of thinking, within the constraints of her own commitments and desires, rather than the substantive value of the premises or outcomes. First, the patient's reasoning satisfies the internalist requirement: the patient's decision to terminate the pregnancy cohered with her subjective pro-attitudes (independent of whether the *content* of those beliefs were extrinsically true or valuable):

> She has said, not only today, but on a number of other recent occasions, that she feels suicidal at the prospect of having to carry this child to term. She says that if there is no termination she will seek to kill herself or the

[24] On top of logical consistency, Richard Brandt in *A Theory of the Good and Right* (Oxford: Clarendon, 1979) adds the further stipulation that an individual's desires are founded on full empirical information – notwithstanding whether their desires *are in fact* grounded in full empirical information. I don't wish to discuss this additional criterion, as it would lead to an over-idealised model that lacks traction with what I see are the relevant assumptions within the 'use and weigh' criterion of the functional test.

[25] Stephen L. Darwall, 'Internalism and Agency', *Philosophical Perspectives* 6, Ethics (1992): 165.

[26] Ibid., p. 165.

> baby. It may be that those suicidal thoughts are in some way bound up with her illness. But if, indeed, she does feel them . . . then it seems to me to be entirely rational for her to consider and decide that it is preferable for her to seek and undergo a termination before being driven to attempting suicide.[27]

Second, her reasoning meets the condition of consistency, where her decision to terminate followed logically from her premises. Applying the transitivity condition, if the patient prefers an abortion to suicide, and suicide to having the baby, then it follows that it is rational to prefer an abortion to having the baby. On the face of it, adherence to the procedural model helps track cognitive deficiencies in logical thinking, without appealing to criteria external to the individual. This in turn ensures judgments of capacity adhere to the core statutory aim of protecting and promoting individual autonomy without passing judgement about the content of their reasoning. Holman J states:

> It seems to me . . . that even if aspects of the decision making are influenced by paranoid thoughts in relation to her husband and her mother, she is nevertheless able to describe, and genuinely holds, a range of rational reasons for her decision. When I say rational, I do not necessarily say they are good reasons, nor do I indicate whether I agree with her decision, for section 1(4) of the Act expressly provides that someone is not to be treated as unable to make a decision simply because it is an unwise decision. It seems to me that this lady has made, and has maintained for an appreciable period of time, a decision. It may be that aspects of her reasons may be skewed by paranoia. There are other reasons which she has and which she has expressed. My own opinion is that it would be a total affront to the autonomy of this patient to conclude that she lacks capacity to the level required to make this decision.[28]

Providing that the norm of consistency is adhered to, contentious or unwise choices can still be judged coherent according to the standards of procedural rationality. Concerns about paternalism and a commitment to liberal value-neutrality make it understandable why legal and clinical attention has focused on the abstract process as opposed to the outcome of reasoning as an indicator of mental capacity.

But though understandable, I want to suggest in what follows that efforts to hive off other, more substantive modes of rational thinking from the functional test of capacity are unsuccessful and may not even be normatively desirable. Two serious problems afflict the procedural

[27] *Re SB*, para. 43. [28] Ibid., para. 44.

model, both which refer to the intersubjective context of reasoning. The first issue surrounds the social function and endorsement of epistemic and practical reasons – or more specifically, the social preconditions without which procedural rationality's preoccupation with formal structure and validity could not get off the ground. The second issue focuses on the social function, disclosure and articulation of practical goods through the use of reason.

II. The Social Space of Reasons

In these next few sections I want to critically examine the assumptions within the model of procedural rationality. First, I want to tackle the acontextual, internalist slant of the consistency condition and replace it with a more holistic account. For the sake of argument, let us accept the assumption that reasoning is indeed coextensive with particular modes of formal thinking, and these modes must adhere to standards of logical validity and justification. When valid inferences are drawn from true or justified beliefs, individuals essentially achieve a certain 'standing' or status in what philosophers (following Sellers) call 'the space of reasons'.[29]

However, a serious objection arises when we consider what is thought to be involved in order to achieve this standing. The procedural model assumes that achievement of this standing is wholly intrinsic to an individual's reasoning process: the independent procedural mechanisms of rationality are themselves set and met internally, based on an individual's subjective motivational set. But this neglects crucial extrinsic, environmental determinants of validity conferral. What results, according to John

[29] The epistemological tradition since Hume has tried to grapple with how we demarcate the 'space of reasons' – namely a sphere of freedom and spontaneity, of conceptual norms and rational justifiability – and the 'space of causes' – namely a sphere of experience, of natural causes and scientific determinism. From one direction, if we reduce everything to the 'space of causes', we effectively deny the spontaneity that is inherent to our cognitive, rational abilities. From the other direction, if we claim that our empirical observations can be separated from the world as it is (the space of causes), it means that our cognitive reflections about the world, our justifications for our empirical beliefs and reasons, have no traction or grip with the world as we perceive and experience it. I don't wish to get into this very important debate in epistemology, though I do want to explain my rationale for continuing to use the term, 'space of reasons'. Although the term might seem ancillary to my normative aim here, Brandom's adaptation of the term seems to me to illustrate helpfully how what counts as *justification* for beliefs and reasons will depend on the intersubjective, discursive environment and context in which we find ourselves. The spatial analogy helps, in so far as we can find ourselves moving from one space to another, and justification of our beliefs and reasons will differ depending on those different spaces.

McDowell, is a 'deformation' and 'interiorization of the space of reasons, a withdrawal of it from the external world. This happens when we suppose that we ought to be able to achieve flawless standings in the space of reasons by our own unaided resources, without needing the world to do us any favors.'[30] Robert Brandom makes a crucial refinement to McDowell: the 'external world' is narrowed to the 'social world' so as to emphasise how the space of reasons is interpersonally and discursively maintained.[31] Brandom's objection is applicable to the procedural model of reasoning in two ways: first, rational justification and validity is conferred through second-personal normative endorsement rather than some intrapersonal process. Second, discursive and interpersonal interaction causally influences the exercise of those cognitive faculties and modes of thought privileged by the procedural model. These two points focus on different junctures of the reasoning process, yet further examination reveals how the social space of reasons is both causally and consequentially implicated throughout.

Brandom's work is especially illuminating for the first point I want to make. According to him, validity and justification are rooted in the concrete social practice of giving and asking for reasons. The space of reasons can be thought of as a 'normative space' that 'is articulated by properties that govern practices of citing one standing as committing or entitling one to another – that is, as a reason *for* another.'[32] Taking responsibility for one's reasons and claims is not enough to achieve a standing in the space of reasons. Rather crucially, validity and justification – or knowledge attribution – depend on the *practical attitude* of a person *receiving another's reasons.* For example, take the reasoning of Anne in the previous chapter: the practical attitudes of the clinicians trying to treat her would not necessarily confer validity on her reasoning for refusing treatment, though it is technically formally valid (i.e. 'Eating makes one gain weight; I value thinness at all costs; therefore I will not eat'). Yet imagine Anne starts to participate in pro-anorexia websites: the practical attitudes she encounters in this forum would likely confer validity and justification to Anne's reasons for not wanting treatment. The procedural model can perhaps acknowledge the initial first-personal conditions of committing to a

[30] John McDowell, 'Knowledge and the Internal', *Philosophy and Phenomenological Research* 55:4 (1995): 877.

[31] Robert B. Brandom, 'Review: Knowledge and the Social Articulation of the Space of Reasons', *Philosophy and Phenomenological Research* 55:4 (1995): 902–3.

[32] Ibid., p. 898, emphasis added.

belief or reason; yet it is the necessary social (second-personal) dimension to justification (where one is bestowed a standing in the space of reasons) that is ignored. When a person is deemed to have true beliefs, justified claims, or valid inferences, this means that a *social standing* is accorded to those beliefs, claims or inferences.

Earlier I mentioned that one of the virtues of the procedural model's normative minimalism is the potential to make an individual's contentious beliefs, reasons, and decisions explicable and coherent. This difference between explicability and endorsement of reasons can be explained using Brandom's distinction between *attribution* and *undertaking*. Three different practical attitudes are involved to achieve the standing of someone who has *true knowledge*: (i) attributing a commitment; (ii) attributing an entitlement; and (iii) undertaking a commitment.[33] Yet all three attitudes need not be present for someone to have justified and understandable belief. For example, my practical attitude could be such that I can *attribute* an individual with a belief that her thoughts are being broadcast over the radio and television; I may also take her to be *entitled* to that commitment, as I see wires from a mechanical device attached to her head. But I might not be able to *undertake or endorse* her commitment since I know that the machine is a scanning device to detect cancerous tumours. Therefore, in our shared space of reasons, I do not accord her justified belief the standing of *truth* in this context – this standing is contingent on another person (myself) undertaking her commitment in the space of reasons. As Brandom notes, such standing is 'essentially a social status, because it incorporates and depends on the social difference of perspective between attributing a commitment (to another) and undertaking a commitment (oneself)'.[34] The different social perspectives involved in attribution and undertaking means that one can attribute a standing to another without necessarily undertaking her commitment: I can accord someone with the status of 'justified belief that is not true'.[35]

Brandom's argument about justified belief applies similarly to the social standing attributed to inferential reasoning. The procedural model may assume that valid logical inferences are non-contextual and can be generated introspectively, yet Brandom has shown that inferential reasoning is itself an expressive, articulatory act that is bound to discursive norms constitutive of social practice – of what is expected when we ask for and

[33] Ibid., pp. 903–4. [34] Ibid., p. 904. [35] Ibid.

provide reasons.[36] Validity is not simply about the *logical form* that an argument or doxastic commitment takes, but can only be conferred if the implicit pattern of inferences can be endorsed *interpersonally*. In short, the social context will determine what counts as good and valid inferential reasoning. Consider the two inferences:

(i) Overeating will give me heartburn, so I will eat until I am just full.
(ii) Eating spicy food will give me heartburn, so I will avoid eating spicy food.

Though not formally valid, these inferences can be deemed *materially* valid, particularly if it expresses a second-personally endorsable preference for avoiding heartburn. Validity in this instance is a reflection of the second-personal normative stance or practical attitude with respect to the implicit, material commitments embedded within the expressed claim. In other words, rather than an abstract, formal, and non-contextual feature that can be adhered to in isolation from others, validity is effectively conferred when the person receiving the inference can *endorse or undertake* the same defeasible reasons that she ascribes to another individual.[37]

My discussion sounds fairly abstract so far. But in practice, Brandom's argument illuminates how disputes about capacity in numerous cases emerge as a result of social disagreement about the status accorded to certain beliefs, reasons, and commitments. It also reveals how, even at the most basic level, the perspective involved in attribution will require a stance of interpretive openness towards another individual, even if her commitment cannot be second-personally endorsed. Applying this framework to three legal cases illustrates its explanatory potential.

I have already suggested one sense in which the judgment in *Re SB* is consistent with the procedural model of reasoning. Deeper probing reveals that the assessment of the patient's reasoning behind her decision appears to hinge on the practical attitude or normative stance of others within the social space of reasons. The judge clearly attributes a commitment and entitlement to the patient's reasons behind her decision to terminate her pregnancy, though he may not go so far as to undertake her commitment. Holman J states, 'she is . . . able to describe, and genuinely

[36] Robert B. Brandom, *Making It Explicit; Reasoning, Representing, and Discursive Commitment* (Cambridge, MA: Harvard University Press, 1994), pp. 141–98, as well as *Articulating Reasons; An Introduction to Inferentialism* (Cambridge, MA: Harvard University Press, 2000), pp. 157–84.

[37] Brandom, *Articulating Reasons*, pp. 90, 168.

holds, a range of rational reasons for her decision. When I say rational, I do not necessarily say they are good reasons, nor do I indicate whether I agree with her decision'.[38] By contrast, her family and the medical professionals do not even attribute entitlement to her beliefs and commitments in the space of reasons, suggesting instead that her 'current persecutory or paranoid beliefs [about her husband and mother are] a result of the bipolar illness', leading to the assessment that 'she is not thinking straight'.[39] From this perspective, the material inferences leading to her decision are invalidated – not because the structure of her reasoning is skewed, but because she is thought to lack entitlement to hold the reasons and commitments she expresses. In other words, she fails to 'act with reasons', making her decision unintelligible to herself and certain others.[40]

This explanatory framework is also revealing when applied to the contradictory medico-legal stance towards competency to refuse of treatment on grounds of religious belief. It is commonly known, for example, that Jehovah's Witnesses do not accept blood transfusions, even if necessary for a medical procedure to save one's life, due to the belief that the Bible prohibits taking blood. However, this religious conviction jars with the medical professional's legal and ethical duty to provide emergency treatment to preserve life and health where possible. Legal cases involving a conflict between these views illustrate well Brandom's point about how inferential validity hinges on whether implicit commitments can be endorsed and undertaken by another.

In Canada, the Ontario Court of Appeal rendered a landmark decision in *Malette v Shulman* [1990],[41] which concerned the question of whether an emergency hospital doctor was to be held legally responsible for wrongfully providing a blood transfusion to an unconscious Jehovah Witness in hypovolemic shock following an automobile accident. Obviously broader, more far-reaching ethical and legal implications arise from this decision,[42] but I want to focus more narrowly on details in the case notes to illustrate how implicit commitments of inferential reasoning are socially and discursively validated. The doctor in question had been aware that the patient held in her a wallet a Jehovah's Witness card, which clearly stated her religious convictions and made the explicit instruction that no

[38] *Re SB*, para. 44. [39] Ibid., para. 32. [40] See Brandom, *Articulating Reasons*, p. 93.

[41] *Malette v Shulman* [1990] (Ont. C.A.) 72 OR (2d) 417.

[42] See Norman Siebrasse, '*Malette v. Shulman*: The Requirement of Consent in Medical Emergencies', *McGill Law Journal* 34 (1989): 1080–98; Barney Sneiderman, 'The Shulman Case and the Right to Refuse Treatment', *Humane Medicine* 7:1 (1991).

blood transfusion was to be given in any circumstances. He administered the transfusions in spite of this, and continued to do so, even after the arrival of, and adamant objections expressed by, the patient's daughter, on grounds that they were 'medically necessary in this potentially life-threatening situation [whereby] he believed it his professional responsibility as the doctor in charge'. The case notes report his scepticism over the validity of the patient's Jehovah's Witness card:

> He was not satisfied that the card signed by Mrs. Malette expressed her current instructions because, on the information he then had, he did not know whether she might have changed her religious beliefs before the accident; whether the card may have been signed because of family or peer pressure; whether at the time she signed the card she was fully informed of the risks of refusal of blood transfusions; or whether, if conscious, she might have changed her mind in the face of medical advice as to her perhaps imminent but avoidable death.

The appellate judge understood the signed card differently:

> Our concern here is with a patient who has chosen in the only way possible to notify doctors and other providers of health care, should she be unconscious or otherwise unable to convey her wishes, that she does not consent to blood transfusions. Her written statement is plainly intended to express her wishes when she is unable to speak for herself. There is no suggestion that she wished to die. Her rejection of blood transfusions is based on the firm belief held by Jehovah's Witnesses, founded on their interpretation of the Scriptures, that the acceptance of blood will result in a forfeiture of their opportunity for resurrection and eternal salvation. The card evidences that 'as one of Jehovah's Witnesses with firm religious convictions' Mrs. Malette is not to be administered blood transfusions 'under any circumstances'; that, while she 'fully realize[s] the implications of this position', she has 'resolutely decided to obey the Bible command'; and that she has no religious objection to 'nonblood alternatives'. In signing and carrying this card Mrs. Malette has made manifest her determination to abide by this fundamental tenet of her faith and refuse blood regardless of the consequences. If her refusal involves a risk of death, then, according to her belief, her death would be necessary to ensure her spiritual life.

On one hand, the patient explicitly expressed religious reasons against receiving a blood transfusion, namely her religious faith proscribes the practice. Underlying the reason is an implicit doxastic commitment that death resulting from obedience to her religion would be preferable to life in such circumstances. On the other hand, the validity of this commitment could be called into question, depending on the social perspective one adopted. The doctor's doubts as to whether the decision against life-saving transfusion reflected her authentic wishes in such extenuating circumstances reveal he could not normatively endorse or undertake the implicit commitments ascribed to the patient. The judge's statements, by contrast, revealed a more amenable practical attitude towards the patient's reasoning. His validation of the patient's doxastic commitment – that death is preferable to life-saving treatment via blood transfusion – does not ascribe a particular property of truth to its actual propositional content, nor does it mean he first-personally embraced the same commitments. One need not go that far. What is more important is how the normative standpoint of the judge assumed that the patient was entitled to her commitments in the space of reasons and should have been treated accordingly.[43]

Consider the similar case of a 14-year-old Jehovah's Witness who was found to lack 'Gillick competency' to refuse life-saving treatment involving plastic surgery and blood transfusions following an extremely severe burning accident.[44] Although the finding was partly because information was withheld from her regarding the full consequences of her refusal – specifically the manner of *how* she would die – the assessment of her competency also clearly illustrates how inferences are interpersonally and discursively validated or invalidated, depending on one's normative attitude towards its underlying commitments. Whilst the patient's reasoning and inferential commitments informing her decision to refuse the necessary blood transfusion had validity amongst her family and church congregation, the practical attitude shared by the judge and consultant child psychologist meant that both components of justification – attribution and undertaking – were withheld. From the standpoint of the latter, the patient's lack of experience beyond her religious faith and community invalidated her refusal. The psychiatrist claimed that the patient's refusal to have a blood transfusion was 'based on a very sincerely, strongly held, religious belief which does not in fact lend itself in her mind to discussion'. Because the view had 'been formed by her in the context of her own family experience and the Jehovah Witness' meetings where they all support[ed] this view', it meant that this could not be an one which reflected the 'constructive formulation of an opinion which occurs with adult experience'. The judge concluded that her 'limited experience of life which she has . . . necessarily limit[ed]

43 See Brandom, *Articulating Reasons*, p. 168.

44 *Re L (medical treatment: Gillick competency)* [1999] 2 FCR 524.

her understanding of matters which are as grave as her own present situation.'[45]

The three examples discussed here illustrate how different intersubjective contexts imbue meaning to the reasons given by another. Even if one insisted that the core norms of procedural rationality function as an appropriate starting point for evaluating the capacity to 'use and weigh' information, we would still need to accept that justification and validity is the product of intersubjective attribution and endorsement of a person's commitments in the space of reasons. In other words, the norms determining valid inferential reasoning are not fixed, context-independent, nor generated and adhered to intrapersonally: they instead reflect the evaluative, practical attitudes of the second-personal, relational environment and can change accordingly.

So far I have been examining how the social space of reasons functions as an ex post facto normative stamp of approval for certain beliefs, claims, and commitments. Importantly, this interpersonal space functions in a circular and reflexive manner, influencing the entire reasoning process, from the development to the explicit articulation of commitments and reasons. The clinical perspective on competency already recognises this point. Consider how the clinical category 'appreciation' is subsumed in the legal understanding of what it means to 'use and weigh' information, whereby many assessments pinpoint a patient's lack of insight, or the presence of delusional beliefs or inflexible, obsessive thoughts. Essentially, what has been compromised is the ability to process claims and commitments that have an existing normative standing in the social space of reasons. Appreciation presupposes what we might call the norm of *adaptability*, which determines whether self-referential beliefs and desires are sufficiently flexible and responsive to commonly held situational and environmental facts.[46] Failure to integrate such facts within the deliberative process suggests rigidity in thinking and an inability to properly process or weigh up new information as it arises.[47] In sum, what

[45] For further discussion, see Jonathan Montgomery, 'Health Care in Multi-Faith Society', in John Murphy, ed., *Ethnic Minorities, Their Families and the Law* (Oxford: Hart, 2000), pp. 161–79.

[46] Bermúdez, 'Normativity and Rationality', p. 468. Bermúdez seems to suggest that appreciation could be divided into 'procedural rationality' (which is guided by formal principles of logical consistency) whilst 'epistemic rationality' denotes a relation between belief and evidence. For my purposes, it makes more sense to make the distinction between procedural and substantive models of reasoning, with the latter invoking norms of the actual content of theoretical or practical reasons.

[47] See Craigie and Coram, 'Irrationality', p. 89.

is being assessed are individuals' attunement to their standing in a particular social space of reasons and their epistemic sensitivity and adaptability to social and environmental truth conditions.

But such attunement, sensitivity, and processing abilities – necessary ingredients to perform the type of consequential and comparative thinking the procedural model of reason privileges – are developed, heightened, and reinforced primarily through socialisation and relationships. Much like how valorisations of rugged individualism in liberal models of autonomy disregard the lived reality of human interconnectedness and interdependence, the procedural model's assumption that cognitive tasks are mainly performed through private introspection neglects the way in which the relational context can influence every juncture of the reasoning process – from the development to explicit expression of thoughts, beliefs, preferences, and commitments.

A personal injury case decided outside the Court of Protection illustrates this point brilliantly. The judgment in *V v R* [2011][48] argued that capacity was *enhanced* precisely because of the decisional support and guidance provided by others. V was a young woman who had been hit by a car and suffered a traumatic head injury as a result. The issue to be decided was whether she had litigation capacity in the subsequent case seeking compensation from the vehicle insurers. Concerns were expressed that V relied too heavily on her parents, particularly her mother, for advice and reassurance. The view of the medical experts was split. One argued that V's main impediment to her weighing information was her impulsiveness, of which could be overcome through relational support: within the structured setting comprising legal advisors and her parents, V had capacity to make appropriate decisions with regards to litigation. By contrast, another expert suggested that these supportive constraints detracted from her capacity, arguing that V was unable to weigh information and would 'in reality be doing no more than agreeing to decisions or agreeing with decisions made by others'.[49] V's mother expressed similar concerns that V would 'just agree to whatever was being proposed or had been advised', should she be left on her own to make decisions without family support. Against the latter, the judge argued, firstly, that the hypothetical scenario envisaged by V's mother was unfounded, and further efforts were needed to try and determine whether V could make her own, true decisions within the constraints of the relational support that was required.[50] Secondly, proper weight had to be accorded to the assistance provided by V's family in her deliberation. He concluded

[48] *V v R* [2011] EWHC 822. [49] Para. 24. [50] Paras. 31–2.

that the claimant would have 'difficulties rather than a straightforward inability to weigh the evidence and make relevant decisions'. He continued, 'those difficulties are capable of being ameliorated, *if not entirely overcome*, by the careful and structured support that the statute contemplates'.[51] Dialogical prompting and interpersonal support was understood as a constitutive factor in V's procedural reasoning and decisional capacities.

What is particularly interesting about the case of V is how her willingness to seek and follow advice from others helped determine both the capacity adjudication for litigation and financial management. In relation to her legal affairs, V stated that 'she would ask for her parents' advice and said to her mother "everything goes through you".'[52] Because the reasoning of others was first-personally endorsed, it was thought to be constitutive of V's own decision-making abilities. Yet the fact that V failed to integrate the reasoning of supportive others in relation to money demonstrated her incapacity to manage her financial affairs. When it came to her finances, V tended to either deceive her parents or disregard their advice. One expert witness tried to encourage her to 'weigh up issues such as whether she needed the money immediately in the context of how much she might have. I tried to get her to think about this but *she said that it would make no difference*'.[53]

This example shows how limitations to one's cognitive skills can fundamentally impede an individual carrying out reasoning tasks alone, making the dialogical prompts and guidance of others crucial in such scenarios. But this is true not only of those with cognitive impairments. Consider the complex interplay between formal reasoning and other faculties, such as emotion, memory, imagination, and volition, in making a practical decision. One view might say that individuals simply probe their own private mental states, self-referential responses and reasons to achieve their aim. This picture, however, is deeply unrealistic – one's motivations and values are often explored through dialogue with others, particularly the more momentous a decision is or more conflicted a person feels. Individuals frequently elicit others to draw logical connections or possible consequences they might not see themselves. Or others might observe how a possible course of action discords with their interpretation of the situation or their understanding of the individual involved. For example, I might say to my good friend, 'If I'm at a work function, remind me not to drink too much'. When she notices I reach for my third glass of wine at a work party, she gently reminds me that I should not drink too

[51] Para. 34, emphasis added. [52] Para. 19. 12.5. [53] Para. 19. 12.4, emphasis added.

much. Perhaps I ask her why, and she explains the reasons and possible consequences – all of which I might have forgotten at the material time. Hence, an individual's attunement to the circumstances and facts of a situation, and indeed, the ability to engage in consequential or comparative thinking, can be shaped by the interpretations of others. Even if one fails to do this prior to making a decision, the explanation after the fact will frequently involve reasoning with others. As Meyers states, '[o]ften . . . people need the additional stimulation of others' interpretations of their situation in order to notice self-referential responses or in order to find plausible explanations of their responses.'[54]

To sum up my discussion so far, I have argued that the implicit acontextualism and internalism of the procedural model is unsustainable because we function within the social space of reasons, both when beliefs and reasons are normatively endorsed, and in the very process of developing those beliefs and reasons. Even if we were to insist that the procedural model's privileging of particular modes of formalistic thinking is indeed correct, we would still have to accept, contrary to internalism, how the social and dialogical context provides the necessary background conditions which develop, affirm, and challenge one's implicit as well as explicit beliefs, inferences, and commitments. As I will ultimately suggest later, this implies that the normativity of reasons – the reasons that have normative force – can originate from external sources.

III. Reason-Responsiveness

In more concrete terms, what I have been trying to get at thus far is how the rational norms implicit in the functional test, such as 'understanding' or 'use and weigh', hinge on the social context to determine which reasons are deemed valid and justifiable, even when working with a procedural understanding of reasoning. This further undermines the law's preoccupation with the individual and one's internal cognition to ascertain whether one has the requisite deliberative, decision-making capacities. Brandom's description of the social space of reasons reveals that procedural standards of consistency are themselves embedded within discursive, intersubjective sphere. It challenges the view that the condition of logical consistency refers (i) to an abstract, acontextual structure where (ii) its successful enactment is specified internally. Recognising the social context of reasoning refocuses our attention, away from subjective conditions of

[54] Meyers, *Self, Society, and Personal Choice*, p. 80.

deliberation and towards the ways that the intersubjective space influences reason's normativity and validation.

On this latter point, however, one might argue that Brandom's account by itself presents no real threat to proceduralist standards of reasoning. The inclusion of 'reason-responsiveness' conditions within proceduralism might accommodate Brandom's description of the social space of reasons and likewise help pick out pertinent examples where external forces influence or manipulate an individual's reasoning (and thereby compromise their autonomy). In this section, I want to focus on the version of reasons-responsiveness John Fischer and Mark Ravizza offer in their influential defense of compatibilism – the claim that free will can coexist alongside causal determinism.[55] Their version of proceduralism posits that factors *external* to the agent *do* matter, particularly if we take a historical (how an individual developed her psychological mechanisms) rather than time-slice view of the agent (what an individual's psychological mechanisms are at this present moment). A time-slice, strict internalist view cannot meaningfully distinguish between the actions and rationality of individuals who act freely, compared with those whose actions are the historical result of mechanisms implanted through manipulation. Fischer and Ravizza seek to make their accounts immune to the latter problem through the introduction of extrinsic, objective standards of reasoning. They stipulate that (i) the processes or mechanisms that individuals act from must be responsive to reasons; and (ii) the individual must take *responsibility for* or *ownership of* the mechanisms sometime during its causal history.

These conditions show that Fischer and Ravizza try to tread the boundary between extreme externalist and internalist positions on rationality. Moderate reasons-responsiveness comprises *reason-receptivity* – the 'capacity to recognize that reasons exist' – as well as *reason-reactivity* – 'the capacity to translate reasons into choices (and then subsequent behavior)'.[56] Though both important, reasons-receptivity has asymmetric importance when assessing the freedom and responsibility of individuals; indeed, individuals must exhibit an 'appropriate pattern of reasons-recognition' so that they demonstrate a recognition of 'how reasons fit together', of 'why one reason is stronger than another', of 'how the acceptance of one reason as sufficient implies that a stronger reason must also

[55] John Martin Fischer and Mark Ravizza, S.J., *Responsibility and Control; A Theory of Moral Responsibility* (Cambridge: Cambridge University Press, 1998).

[56] Ibid., p. 69.

be sufficient'.[57] Particularly reminiscent to the third-party task of judging an individual's reasoning abilities implicit in mental capacity assessment, they write:

> On our approach, it is as if a "third party" (the one assessing the moral responsibility of the relevant agent) conducts an "imaginary interview" with the agent. In this interview, he asks about various actual and hypothetical scenarios, and elicits views from the agent as to what would constitute sufficient reasons. Even if a person claimed that, given his actual values (or preferences), only one reason counts as sufficient, the pattern of his actual mechanism's receptiveness could still be tested by asking him which reasons would count as sufficient, given *another* set of values (or preferences). The third party then employs the information from the interview, together with background information, to seek to understand the pattern in the set of reasons-recognitions. What is required is that the configuration of answers in the imaginary interview can (together with background information) give rise to an understandable pattern, from the perspective of the third party (the person judging whether the agent is morally responsible).[58]

We can imagine a capacity interview going along these lines precisely, where the assessor evaluates the pattern of an individual's reasons-receptivity to determine whether certain 'objective' conditions have been met. These objective conditions would be satisfied through evidence that gradings or relative strengths have been attributed to reasons that render the individual's pattern comprehensible (similar to the transitivity condition) but, by the same token, demonstrate some understanding of the world – where their reasons-receptivity is grounded in some grasp of external, empirical reality.[59] Conversely, the more moderate internalist dimension of this account emerges in the condition that individuals must subjectively recognise, take ownership and responsibility, of mechanisms that cause them to act.[60] This version of procedural reasoning thus requires that we see ourselves as agents who are receptive to reasons within our community, where we display sensitivity to the reactive attitudes of others in the formation of beliefs and reasons.

Examples might help here. Consider the charity example: Tom might recognise certain reasons for giving to charity – he can see that it is of relatively small cost to himself for the benefit of others, that it is an objectively good and ethical thing to do, and that he might gain some

[57] Ibid., pp. 70–1. [58] Ibid., pp. 71–2. [59] Ibid., pp. 72–3.

[60] In calling this 'internalist', I am referring to the usage in Williams: namely how normative reasons must latch onto an individual's subjective motivational set to be truly one's own reasons.

praise from his friends. But Tom still resists giving to charity, choosing to spend money on silly items for himself. When asked to justify his reasons, he says, 'I recognise that all of these reasons (costs, benefits, praise, etc.) might be sufficient for one to give to charity, but I still choose otherwise'. This shows a certain degree of reasons-receptivity, even as he might lack reason-reactivity, in that Tom makes a divergent choice even as he recognises the strength of these reasons when pushed to justify his actions. Or consider an example where reasons-receptivity might be lacking: imagine Rob (from the previous chapter) experiences auditory hallucinations as a result of his schizophrenia. He says the only reason for him to take his medication is 'when the Virgin Mary tells him to', meaning that there are times he doesn't take it. Others try to explain alternative, sufficient reasons for him to take his medication, such as for his physical and mental health, or might appeal to how it helps other things that matter in his life, such as his relationship with mother. Yet he dismisses these, saying that 'none of these are reasons to take my medication. It's only a reason when the Virgin Mary instructs me.' In this case, Rob might be thought to lack reasons-receptivity: he cannot understand how the alternative reasons might be sufficient for him to take his medication, even when instructed otherwise by the Virgin Mary.

Fischer and Ravizza's account of proceduralism is important, in that they appeal to objectivist, external criteria which require individuals to engage with the realities of their environment, such as the affirming or blaming attitudes and the held facts and values of their intersubjective community. Indeed, one could say that this version of proceduralism works in concert with Brandom's holism: they provide the broader, metaethical justification for 'objective' criteria of reasons-responsiveness to cite as evidence when individuals *lack* procedural rationality (and thus maintain its commitment to content-neutrality), whilst Brandom shows in detail the complex reciprocity between individuals' mechanisms or patterns of inferential reasoning and their intersubjective environment. If this is correct, it shows that proceduralism need not be strongly committed to internalism (and can indeed accommodate externalist criteria of reasoning), thus potentially undermining the argument I have mounted thus far against the procedural model of rationality, particularly as interpreted in the functional test of capacity.

But I believe that proceduralism of this stripe still encounters problems to do with its content-neutrality. Recall that one of the redeeming virtues of proceduralism is its alleged neutral stance towards the substantive content of reasons: whether these reasons are 'good', 'bad', 'wise' etc., is set aside, and the formal structure of one's reasoning remains the focus.

Fischer and Ravizza use an example of where an individual's pattern of reasons-recognition has gone wrong: consider Jennifer who would not attend a basketball game if the tickets cost $1,000, yet would go if the tickets cost $2,000, or every other scenario, i.e. if it cost $5, $50, or $999, so long as it didn't cost $1,000. The pattern of reason-recognition appears ad hoc, arbitrary, inexplicable, indicating that something is awry with her reasons-responsiveness mechanism. Similar issues seem to afflict Rob who takes his medication only when the Virgin Mary instructs him to. Indeed, focusing on the structure rather than content of reasons can potentially track what is at issue here. But consider the case of Joan I discussed in the previous chapter, the dementia resident who refuses to see her family but chooses to start a relationship with another resident, contrary to all her long-standing, previously held values and preferences. Suppose that delusional beliefs that her husband of fifty years is now seeing someone behind her back motivate her rejection of her family. Fischer and Ravizza's proceduralism could respond in two, equally indecisive ways:

(i) Joan's sharp change in behaviour would indicate an incomprehensible pattern of reasons-recognition, whilst her delusions demonstrate an incapacitous reasons-receptivity. Whether she has come to those beliefs through her own process, rather than some psychological mechanisms outside her causal control, is doubtful.
(ii) Joan's behaviour meets the criteria of procedural rationality given that her reasons (comprising both deluded and empirically grounded beliefs) convey a sufficiently comprehensible pattern of reasons. Though she has dementia, she subjectively takes ownership for these reasons and recognises herself as a responsible agent, accepting her family's sense of anger towards her. When she is confronted that her beliefs about her husband are untrue and likely the result of her dementia, she responds that the dementia is part of her now, and this does not change her mind about cutting off her family.[61]

It seems to me that, even with the introduction of objective criteria, Fischer and Ravizza cannot truly help us assess Joan's reasoning in this case.

[61] Fischer and Ravizza cite three conditions for taking responsibility: (i) the individual must 'see himself as the source of his behavior' or 'see himself as an agent'; (ii) the individual 'must accept that he is a fair target of the reactive attitudes as a result of how he exercises this agency in certain contexts'; and (iii) the individual's 'view of himself as an agent and sometimes appropriately subject to the reactive attitudes be grounded in his *evidence* for these beliefs' (ibid., pp. 210, 211, 213).

If (i) is the right assessment, this raises questions as to whether their objective criteria of procedural rationality is overly demanding. Indeed, if we are to focus on the question of moral responsibility, this reading might have its benefits, in that Joan would have reduced or no culpability for her actions. But as a standard of reasoning to help in the task of capacity assessment, the objective criteria could potentially deem many individuals with weak reasons-responsiveness, with psychological mechanisms like impairments and disorders, as irrational – and thereby incapacitous. If (ii) is right, then Fischer and Ravizza overly prioritise the formal, internalist biases of other proceduralist models and become vulnerable to the criticisms discussed in the earlier section. The fact that Joan might take responsibility for the mechanism may lead to a judgement that she is acting with sufficient reasons-responsiveness, insofar as she takes responsibility for her actions and acknowledges herself as appropriately subject to the reactive attitudes of others (i.e. anger from her family). However, such criteria fail to evaluate disconnections between her previous and current reasons-responsive mechanisms, or how her current (internally consistent) mechanism might be the result of manipulation that, though an intrinsic part of her, she cannot be responsible for.

But let's assume that Fischer and Ravizza would claim that a proper assessment of Joan's rationality might fall somewhere between the fuzzy boundaries of 'rational' and 'irrational'. Indeed, they suggest the softened boundaries between the two are important, particularly if we want to capture the complex, indeterminate nature of free agency. They claim that a key virtue of this more objectivist version of procedural rationality is that it can deal properly with the problem of manipulation. Emphasis on the historical process by which one acquires one's mechanisms means that they can allegedly resist conclusions that someone who is manipulated through some artificial means is rational, free, and morally responsible. My intuition is that Fischer and Ravizza's assessment of Joan's scenario would likewise lean towards (i), particularly as they specify that individuals must not acquire their beliefs through some sort of reasons-responsiveness mechanisms that are explicitly out of an individual's control, such as manipulation or brainwashing. Pathological mechanisms and cognitive impairments would surely fit within the same category. Conversely, if they state otherwise – that it is whether individuals take ownership of these pathological but reasons-responsive mechanisms – this likewise seems counterintuitive: to say that 'Joan is procedurally rational because she understands that she chooses and acts in accordance with her dementia' just seems odd.

In other words, Fischer and Ravizza do not adequately address scenarios where the source of manipulation for one's reasons responsiveness can itself be seen as part of oneself, tracking one's internal reasons, in a wholly egosyntonic manner.[62] Consider Anne's scenario from the previous chapter. Here we have a case where her reasoning to refuse treatment might very well meet the proceduralist criteria Fischer and Ravizza set out. She understands the historical genealogy of her reasons formation; her decisions display a strong, coherent pattern of reasons-responsiveness. Suppose that she takes ownership of the mechanism controlling her and also that she can be on the receiving end of reactive attitudes. She *understands* the reactive attitudes of others though she doesn't respond as they wish, given that these are her subjectively endorsed commitments.[63] In short, she sees herself as a responsible agent even as, from another perspective, one might think that she is being manipulated by her anorexia nervosa, which, going further back, toxic relational and social conditions brought about. According to Fischer and Ravizza's analysis, it would seem that Anne is appropriately reasons-responsive: though she recognises the *originally alien* source of manipulation (her eating disorder or her relational influences, depending on how far we go in the causal history), she takes her eating disorder to be *part of herself*. She might reiterate an attitude of herself as a responsible, free agent (and therefore meets the conditions of moral responsibility). But the proceduralist criteria does not allow us to make appropriate evaluative judgements in cases like Anne's – where the manipulation has such an egosyntonic structure and the individual acts on a mechanism that is strongly responsive to reasons which track their own attitudes about themselves.

Fischer and Ravizza outline an 'appropriateness' condition that is meant to ensure those who have been 'electronically induced to have the relevant view of himself' (and thereby fulfil partial conditions of taking responsibility) 'has *not* formed his view of himself in the appropriate way'.[64] This condition remains unspecific and does not rule out causal determinism, whereby external circumstances determine our belief (though this would be slightly more problematic in the case of

[62] Cf. Steven Lukes, *Power: A Radical View*, 2nd ed. (Basingstoke and New York: Palgrave Macmillan, 2005).

[63] Cf. Stephen Darwall, *The Second Person Standpoint: Morality, Respect, and Accountability* (Cambridge, MA: Harvard University Press, 2006) and my challenge to Darwall's theory in 'The Space between Second-Personal Respect and Rational Care'.

[64] Fischer and Ravizza, *Responsibility and Control*, p. 236.

our motivational states).[65] Ultimately, however, the condition of appropriateness provides insufficient content to determine those cases where external causal determinants violate reasons-responsiveness, and those cases where it does not, particularly where individuals claim ownership of their reasons-responsiveness mechanism even with the recognition that its genealogical origin was alien to themselves.[66] It ultimately comes down to individuals' sense of themselves. But there are many troubling instances where individuals adopt mechanisms as their own (such as in consented to, abusive relationships), taking responsibility for them in the manner Fischer and Ravizza's proceduralism requires, but where we might question whether they should be taking responsibility for these mechanisms (from the first-person perspective), or whether we should be judging them as responsible for those mechanisms (from an agent-neutral standpoint).[67]

Thus, whilst this reasons-responsiveness version of proceduralism might accommodate Brandom's holism about the social space of reasons somewhat, it ultimately leads to counterintuitive conclusions, where individuals manipulated through some biological mechanism or through relational circumstances, could be deemed rational and responsible. This is because at root, the standard of procedural rationality still remains subjectively driven, committed to the internalist condition, where its criteria are still strongly 'relativized to the inclinations of particular agents'.[68] Yet reasoning is not an interiorised, introspective process, nor is its criteria completely abstract and subjectively driven. Although a simple and fairly obvious explanatory point to make, a deeper normative issue is at stake. Recognising the inescapable social dimension to even this most deflationary account of reasoning places significant pressure on articulating the *quality and character* of the social space of reasons – of making discursive and practical attitudes explicit, particularly in light of their effect on individuals' deliberative abilities. In *V v R* the character of V's relationships and how advice was provided to her helped mitigate her suggestibility and impulsivity. Given different circumstances, these characteristics could have been exacerbated. If the different skills and faculties

[65] Ibid., p. 236.

[66] See Eleonore Stump, 'Control and Causal Determinism', in S. Buss and L. Overton, eds., *Contours of Agency: Essays on Themes from Harry Frankfurt* (Cambridge, MA: MIT, 2002), pp. 33–60.

[67] See my paper, 'Space between Second-Personal Respect and Rational Care' for more on this subtlety.

[68] J. David Velleman, 'The Possibility of Practical Reason', *Ethics* 106:4 (1996): 704.

involved in reasoning can be *supported*, they can equally be *manipulated*, depending on the relational and dialogical context. Fischer and Ravizza's brand of proceduralism doesn't help us make this distinction where it counts.

Yet one might argue that recognising the *social* nature of the space of reasons doesn't help us along with our task any further either. Brandom tries to describe the manner that the social space of reasons functions but ultimately falls short of making normative claims about the substantive nature of this dialogue. The distinction between relational support and manipulation may not be clarified any better on applying his account. I don't think this is the case, however. In the first place, Brandom's theory is important because it highlights the unavoidably interactive, malleable, and variable nature of the social space of reasons. This is something to be accepted and embraced. Acknowledging that the medico-legal sphere is itself composed of socially articulated reasons that are context-dependent, and will develop and change through discursive engagement, calls into question the assumption that certain reasons have automatic epistemic or practical warrant. Greater critical reflexivity about paternalistic and perfectionist judgements can be encouraged.

If my argument is correct thus far, it seems to me that any model of reasoning that leaves us inarticulate about value has to be abandoned. Acknowledging the social, intersubjective nature of the space of reasons makes it vital that we move away from the content-neutrality of proceduralism, towards an account that allows us to (i) become explicitly aware of and articulate the implicit commitments informing our intersubjective spaces; and (ii) make critical judgements of a more substantive sort. I turn my attention to this in the next section.

IV. Value-Constituting Frameworks and Reasoning

As we see in the strong rejection of outcome-based and status-based tests of mental capacity, there is an aversion to allowing third-party assessments of the content of individuals' reasons, due to concerns about paternalism. Ideally, the test of mental capacity is meant to be neutral towards a person's reasons, values, and ends. A key merit of assuming the procedural model of reasoning in functional capacity tests is its value-neutrality, given that its evaluative focus is on the formal structure with intrinsically generated norms, as opposed to substantive content of reasoning. Worries that unwarranted paternalistic judgements about the good life are being imposed on capacitous, self-determining individuals can then be assuaged.

But whether capacity assessments can remain value-neutral even when using a procedural approach is dubious. At times assessments of 'use and weigh' appear to rest on an implicit regress strategy, beginning with the assessor's value-loaded question, 'would a reasonable person (or I) choose the same way?' Sometimes it's not just implicit. For example, one expert medical witness in *Loughlin v Singh & Ors* [2013] summed up his assessment of the Claimant's capacity the following way: 'The critical thing for me is: does his mind present to him alternatives. And I rather think that he doesn't see alternatives in the same way I do.'[69] I am not saying that all applications of the functional test are like this. But the extreme cases often alert us to problems that may otherwise remain hidden. Failure to realise the ideal of value-neutrality is not only understandable in light of the often intractable nature of the ethical dilemmas involved in capacity adjudications, it would seem more realistic to acknowledge how assessments of 'use and weigh' *must* draw upon judgements of a more substantive kind. The space of reasons is not just social, but also fundamentally value-laden in nature, as it comprises unavoidable, socially affirmed commitments. If this is correct, we need to ask *what* commitments have an existing normative standing and *why*? The answer requires one to articulate the type of community situating an individual, as well as to evaluate the values and commitments constitutive of our accepted interpersonal reasons. Our normative focus would then shift to the pressing issue of how we can make evaluative judgements *better*, heightening the urgency for a mode of thinking that the procedural model occludes. It will require a more substantive form of rationality that can disclose, articulate, and assess the value of certain reasons and goods.

Whilst thought to be more characteristic of best interests decision-making, this type of reasoning is already at play in numerous capacity adjudications. Consider the issue of whether capacity assessments should privilege an episodic or programmatic approach towards the norm of consistency. As a core principle of procedural rationality, this norm can be applied to an individual's programmatic, diachronic choices (with the assumption that an agent's reasons, beliefs, preferences, and desires remain relatively stable and coherent over time) or her episodic, synchronic choices (with the assumption that this stability and coherence applies for a discrete temporal period). Yet when an individual's

[69] *Loughlin v Singh & Ors* [2013] EWHC 1641 (QB), para. 34.

(equally consistent) episodic and programmatic reasoning pulls in opposite directions, the normative minimalism of the procedural model means that which ought to have priority cannot be determined. Another tack would be to use the statute itself to arbitrate such conflict. The decision- and time-specific emphasis of the functional test implies the adoption of a synchronic perspective towards consistency when assessing capacity, whilst the determination of best interests requires a more long-range, programmatic approach. In reality, where there is conflict between episodic and programmatic consistency, legal judgments have tended to prioritise the latter approach towards reasoning and practical agency. More weighty evaluative judgements, rather than the norms of procedural rationality or literal interpretation of the MCA, shoulder the main justificatory burden for these capacity adjudications. For example, we can see how, on one hand, the patient's decision to have an abortion in *Re SB* was rationally inconsistent from a synchronic perspective as it directly contradicted her expressed beliefs, reasons, and choices at earlier stages of the pregnancy. On the other hand, the decision was consistent with her overall long-term values, especially considering her growing doubts about her ability to raise the child within the context of her lifelong, relapsing bipolar disorder.[70] This and the fact that her choice was diachronically consistent with a previous choice made in similar circumstances underlined Holman J's judgment.

Similarly, inadequacies in the patient's programmatic planning led to a finding of incapacity in *Loughlin v Singh & Ors* [2013]. The case surrounded a finely balanced debate concerning the Claimant's capacity to, firstly, conduct litigation for an assessment of damages, and secondly, manage his affairs and property. Following an accident twelve years previous, Mr Loughlin had suffered a severe brain injury, which led to impairments in motivation, initiation, and organisation. His cognitive abilities could, however, be enabled through structured support. On those grounds, expert witnesses on behalf of the Defendant argued that Mr Loughlin possessed capacity in the areas concerned, despite concerns over his habits of impulsive buying as well as sleeping and rising late.

But the testimonies of the expert witnesses on behalf of the Claimant were particularly illuminating for my purposes. They argued that Mr Loughlin's difficulties in structuring and organising his activities were indicative of incapacity, despite his strong performance on tests of executive function. One claimed that lack of structure and prompting meant

[70] *Re SB*, para. 42.

that his lifestyle would be 'chaotic, risky and unproductive'.[71] The absence of long-term planning on Mr Loughlin's part meant that,

> Should he be in charge of a large amount of money, he's going to have great difficulty formulating the problem that he has, because the problem that he has is something to do with making this last. He has asked his deputy whether he can spend the first bit on holidays . . . I think it's to encompass all the decisions he has to make and also to be aware of what the consequences are of failing to make a decision.[72]

Another expert witness pinpointed problems with assessing capacity through the cognitivist, value-neutral lens of the procedural model. In addition to the ability to construct a long-range, programmatic life plan, non-cognitive factors such as emotion were stressed as key constituents of decisional capacity:

> I think the particular difficulty he has [is] in weighing up the consequences of his decision. From reading Mr Loughlin's behaviour, he appears to be very much in the present. He operates in the present tense. And he doesn't seem to be able to anticipate the consequences of his actions either at a behavioural or an emotional level. Therefore I think that he's very much disadvantaged in terms of his decision-making processes was to what the short, medium, or certainly, long term consequences of his decision would be. The other problem is – and it goes back to the emotional aspect – the way the Mental Capacity Act is constructed gives the impression that the cognitive or intellectual side of things is the most important, in terms of language, memory and rationalising your decision. But, in fact, emotion plays a very powerful part in our decision-making. And my impression is that Mr Loughlin's emotional understanding of the world and people has been impaired.[73]

The judge ultimately concluded in favour of the expert witnesses on behalf of the Claimant. He suggested that, whilst vulnerability to exploitation or propensity to make rash, irresponsible decisions need not necessarily suggest incapacity, such contextual matters could and should be taken into account in the capacity assessment.

Yet it isn't entirely clear that Mr Loughlin lacked capacity if the third pillar of the functional test was interpreted in accordance with procedural model standards. The reasoning behind his decisions to use his money unwisely on vacations and guitars, or sleep and rise late, could satisfy the norm of consistency from a synchronic perspective even as it might demonstrate a lack of long-term prudence, wisdom, or discipline.

[71] *Loughlin v Singh & Ors* [2013], para. 33, 9.1. [72] Para. 35. [73] Para. 36.

Judgements of the latter sort, however, violate the procedural model's commitment to value-neutrality. That the expert witnesses were conflicted about Mr Loughlin's capacity is revealing. On one hand, expert witnesses on behalf of the Defendant applied the functional test in a way that reflected the procedural model's commitment to value-neutrality. On the other hand, more substantive judgements grounded the capacity assessment carried out on behalf of the Claimant – such as the importance of making prudent, long-term plans to safeguard his savings and the need to demonstrate emotional aptitude in decision-making. In short, the latter assessment was grounded on judgements about whether Mr Loughlin's values were thought to be appropriate in the situational context. The judge's agreement with the Claimant's expert witnesses suggested that, when assessing capacity, the evaluation of an individual's programmatic choices about the good life may be warranted in certain circumstances, even if episodic, unwise choices satisfy the criteria of procedural rationality.

One could respond that this still doesn't make a substantive model of reasoning justifiable or normatively desirable: the phenomena may very well reveal that more value-laden judgements have informed the practice of capacity adjudications, but it could still be an illegitimate move. I disagree. Brandom's explanatory insight reveals how values and commitments are continually affirmed or discredited – not just in the epistemic domain, but also in the social space of *practical* reasons. Clearly, the procedural model cannot stand alone once we recognise both the intrinsically social nature of the space of reasons and expressive function of rational deliberation. Relying on a procedural model of rationality means we remain inchoate about *what* those substantive goods and values are and how they influence our agency. I am not saying that the procedural model has *no* place in the functional test but that we need to rethink the function of reasoning and its presumed content- and value-neutrality. This is especially so when capacity seems to hinge on the support provided by a person's relationships. I argue that such scenarios instigate a more substantive, *articulative-disclosive mode of reasoning*[74] – of which frequently lurks in the background yet remains unacknowledged. This mode of reasoning is 'where we use language, or some symbolic form to articulate and thus make accessible to us something – a feeling, a way of being, a possible meaning of things – without making any assertion

[74] Charles Taylor, *Dilemmas and Connections* (Cambridge, MA: Harvard University Press, 2011), pp. 39–55.

at all'.[75] Crucially, the articulatory-disclosive mode of reasoning engages one's implicit commitments and understandings, recognising the significant role of background comparative judgements that do not actually appeal to the criteria of procedural reasoning for its validity or force.[76]

Here is likely where I take an additional step of which Brandom just falls short. On one hand, the social space of reasons – the 'package deal' of making claims, the giving and asking for reasons – applies for both fact-establishing (epistemic) and practical (action-guiding) reasons. Acceptance, validation of reasons, cannot exist as an autonomous activity, removed from other symbolic forms of meaning, and Brandom himself admits his rationalism isn't committed to hiving off these other (non-rational) dimensions that may very well capture something vital and inescapable about human life.[77] Yet Brandom remains relatively ambivalent about the necessity of disclosing and evaluating the broader symbolic forms and substantive reasons that comprise our value judgements.[78] But this seems to me a natural extension of Brandom's account. In recognising that the validation of reasons lies with sources external to the self, Brandom is already in territory beyond the internalist, content-neutral commitments of proceduralism. Or to put in Velleman's words,

> [T]he externalist cannot indefinitely postpone giving substantive characterizations of rationality or reasons. The externalist must at some point provide practical reasoning with a substantive standard of success, which will either consist in or give rise to a substantive account of the features that constitute reasons for an action. The externalist will then have to justify his normative judgment that an agent ought to be swayed by consideration of the specified features.[79]

This problem has motivated philosophers, like Charles Taylor, to go further than Brandom in contesting the assumptions behind the procedural model, recognising that the social space of reasons requires the intersubjective disclosure, expression, and validation of certain

[75] Ibid., p. 42.

[76] Charles Taylor, 'Explanation and Practical Reason', *Philosophical Arguments* (Cambridge, MA: Harvard University Press, 1995), pp. 34–60.

[77] See Taylor, *Dilemmas and Connections*, pp. 39–55, also Robert Brandom, 'Reply to Charles Taylor's "Language Not Mysterious?"' in Bernhard Weiss and Jeremy Wanderer, *Reading Brandom on Making It Explicit* (Abingdon: Routledge, 2010), p. 303.

[78] 'I do not claim that the "factional-practical is self-sufficient", except in the sense that I can give sufficient conditions for what one is doing to be doing that without having to mention symbolic forms in a wider sense (even if in fact you couldn't have one without the other).' Brandom, 'Reply to Taylor', p. 303.

[79] Velleman, 'Possibility of Practical Reason', p. 703.

practical commitments. Much like our epistemic commitments, those in the practical domain comprise qualitative distinctions and are not only unavoidable, but they effectively function as situating frameworks or orientating horizons that confer worth and meaning to one's judgements and choices.[80] Certain spaces judge specific reasons, goods, and moral intuitions better and more significant. Those with uncommon power and depth are animated by what Taylor calls *hypergoods*[81] and exercise a powerful influence on us accordingly; they transcend subjective preferences and 'stand independent of our own desires, inclinations, or choices, [as] they represent standards by which these desires and choices are judged.'[82] For example, think of what lies behind the Jehovah's Witness's refusal of blood transfusion: hypergoods, such as religious faith, divine will, or connection with God, animate and reframe all other goods this individual might hold, leading to a reorientation of practical options when deciding about clinical treatments, or even a re-evaluation of the value of her life. Other hypergoods Taylor cites include happiness, justice, equal respect, self-fulfilment, and the affirmation of everyday life.[83]

We can remain uncommitted to which hypergoods are worthwhile and still come away with two important points. First, it is fairly obvious that deeply resonant practical values (such as autonomy, the right to life, the promotion of health and well-being) function as an implicit orientating framework within mental capacity law. What it means to promote capacity will similarly affirm certain constitutive values. Second, the inescapable nature of our value frameworks has significant implications on what we think the function of reasoning is: it cannot be simply about the validation of logical connections between propositions, nor the prescription of what one ought to do based on an agent's isolated, self-referential responses in order to satisfy the norm of internal consistency. In other words, even as applications of the functional test appeal explicitly to the model of procedural rationality with the aim of remaining neutral towards the content of one's reasons, a background understanding of shared hypergoods – of socially affirmed value commitments and practical judgements – nonetheless situate and guide its application. Certain significant goods, like the good of autonomy, of life, of the importance of social belonging, are implicitly adhered to, yet the procedural model

[80] Taylor, *The Sources of the Self* pp. 20–6, 33–5.
[81] Ibid., p. 4. [82] Ibid., p. 19. [83] Ibid., p. 65.

of reason leaves us inarticulate about their value and meaning, about the deeper reasons behind the determinants of action.[84]

By contrast, accepting the presence of orientating value frameworks makes a type of articulatory-disclosive rationality necessary. Cases where it is an incommensurable choice between life and self-determination highlight this point well. In *A Local Authority v E & Ors* [2012][85] Jackson J made a decision to impose life-preserving treatment on a severely anorexic woman who had already begun a palliative care pathway. The judgment was deeply controversial for a number of reasons,[86] but what I want to note here is his striking description of the decision facing him. 'At its simplest, the balance to be struck places the value of E's life in one scale and the value of her personal independence in the other, with these transcendent factors weighed in the light of the reality of her actual situation.'[87] Such a statement suggests that profoundly difficult medico-juridical decisions must reference how values and goods are situated within our moral framework. Deciding between what appear to be equally powerful goods demands the articulative-disclosive mode of reasoning, which allows us to express to ourselves as well as others the conception of goodness and value that bears on our decision-making, moving us in our lives. Reasoning in this sense involves the explanation, clarification, and elaboration of our surrounding implicit moral landscape and of the qualitative values reflected there.

Yet, as noted in my earlier discussion of epistemic beliefs and reasons, strong evaluation of reasons and goods is not carried in isolation. The core essence of this articulatory-disclosive model of rationality is its social, discursive, and relational quality; our situating values might motivate us implicitly, but through language and dialogue, they are disclosed explicitly. Background value judgements about the good or flourishing and the relational dimensions of decision-making are deeply intertwined. In other words, practical choices regarding one's good are rarely purely self-referential or self-generated; rather, they are part of a broader network of significance and meaning with commitments, desires, and motivating reasons that are guided by, responsive to, and resonate with our social, relational environment.

[84] Ibid., p. 89. [85] *A Local Authority v E & Ors* [2012] EWHC 1639 (COP).

[86] See Kong, 'Beyond the Balancing Scales', where I analyse in depth the problems with both the capacity and best interests assessment.

[87] *A Local Authority v E & Ors* [2012], para. 118.

Let me get to the crux of my discussion so far. I have been establishing that social practice and the dialogical context is fundamental to not just the *epistemic formal* norms guiding our inferential articulations, but the *practical* values guiding one's decision-making.[88] The space of reasons in terms of both epistemic and practical justification functions in a circular manner: a discursive, social space normatively endorses one's reasons and commitments; by the same token, reasons and commitments are generated and shaped through the value-laden, relational framework situating the individual. If this is indeed how the space of reasons functions, the case for the articulatory-disclosive model of reasoning becomes stronger, as *critical intersubjective reflection of these spaces can only emerge once an articulatory-disclosive rather than procedural mode of reasoning is deployed within this circle.* A procedural approach leaves us comparatively inchoate about the actual frameworks of value situating our decisions: proper understanding and dialogue about the goods and values we may or may not share is possible only once we embrace reason's expressive and discursive function.

The issue of whether to compulsorily treat eating disorders is a good example here. Consider the case of Anne from the previous chapter and imagine her eating disorder has progressed to the point where clinical professionals have to determine whether she has capacity to refuse treatment. Based on strict procedural norms, her reasoning is internally consistent and valid. But the determination of her capacity to refuse might very well hinge on the *content* of the values motivating her decision, and the care team's assessment of these, based on their own moral frameworks.[89] If Anne states, 'I don't want treatment because I value thinness above all else', it is more likely that disagreement about her capacity may very well lie in the debatable quality of her value. If Anne states, 'I don't want treatment because I value my autonomy above all else', the discussion of her capacity might alter slightly – mainly because there is an acknowledgement that this is a shared good that has some resonant force. Her care team might debate about what autonomy really is in this case, but that is the *precise point*: that to discuss and reach a conclusion about Anne's capacity, they have had to move well beyond the confines of the procedural model, and engage in a more articulatory-disclosive mode of reasoning.

[88] For further analysis of this, see especially Brandom, *Making It Explicit*, pp. 199–270.

[89] Tony Hope et al., 'Anorexia Nervosa and the Language of Authenticity', *Hastings Center Report* 41:6 (2011): 19–29.

In making this argument, one might accuse me of making an unjustified slide from a discussion of the model of reasoning ability the *patient herself* needs to exhibit (first-personally) in order to possess decisional capacity, to the model of reasoning used and applied by *medico-juridical* practitioners when they assess the patient. Whose reasoning am I talking about? And if we accept that a more substantive, articulatory-disclosive model of reasoning operates at *both* levels, does this in turn raise the standard of reasoning a patient has to attain to demonstrate decisional capacity? This is not only problematic in the sense that the requisite level of articulacy about certain reasons and goods would appear too demanding, it would also allow paternalism to enter through the backdoor. Neither prospect is desirable.

I think that my discussion of open dialogical conditions and hermeneutical practice in Chapter 6 can assuage these worries. For now, let me defend my account in a slightly indirect manner: how and why the articulatory-disclosive model applies to both the assessor's and patient's reasoning becomes clearer once we understand the interaction between two seemingly separate, yet interrelated, spaces of reasons, particularly when the capacity of the individual is disputed. In these circumstances, assessments of capacity invoke first- and second-order spaces of reasons: the first-order context – what we might call the *personal-supportive* space of reasons – orientates the individual in her daily life and care. This refers to not only or necessarily the family, but the more intimate care and relational context of the individual. The second-order context is the *medico-legal* space of reasons, which shapes a capacity assessor's epistemic and practical normative attitude in his or her discursive engagement with a patient. To clarify, these terms are mere placeholders, and the hierarchical structure between them is not normative but descriptive. An implicit power dynamic structures the interaction between the two spaces of reasons, particularly in cases where capacity is contested or an assessment is triggered by suspicions of neglect, abuse, or oppression in the immediate relational environment of individuals with impairment. In such scenarios, judgments of the second-order context tend to be authoritative and legally binding.

These different spaces of reasons are nonetheless dynamic, are not strictly separate, and can share overlapping reasons. Nor is its hierarchical structure set in stone, for the power dynamic that emerges in situations where capacity is contested can nonetheless be mitigated through sensitive dialogical practices (as discussed in Chapter 6). But for my present purposes, the more important normative point is the central role of the articulatory-disclosive mode of reasoning when these two spheres intersect via the individual whose capacity is being assessed – or more specifically, her self-conception and self-reading that is disclosed through articulation. Articulated self-readings make explicit an individual's epistemic and practical commitments as well as self-referential responses. It also discloses how a broader relational framework constituted by reasons, value, and goods fosters and shapes their particular self-understanding. Crucially, this self-reading can be interpreted and assessed from both first- and second-order perspectives. My claim is that, if the second-order, medico-legal sphere passes normative judgement on whether an individual's relationships support capacity, then *an individual's expressed self-understanding and self-reading should be the point of entry for assessing the space of reasons at the first-order, personal-supportive level.* Scrutiny of an individual's self-reading and self-understanding will reveal whether either *both* or *neither* of her immediate relationships/primary care team and capacity assessors actually encourage the development of key autonomy competencies that were discussed in the previous chapter. In this respect, my account can help us with the task of explaining how capacity adjudications can distinguish between those relationships that detract or enhance an individual's decisional capacity without necessarily sanctioning a paternalistic approach. How this is so will become even clearer in the following section.

V. Application

Thus far my discussion of articulatory-disclosive reasoning has been fairly abstract. To be clear, I am not suggesting that all of these philosophical assumptions do in actual fact inform the reasoning behind the judicial decisions. But it does provide an illuminating normative prism through which to interpret cases in which the relational context is a crucial determinant of individuals' capacity. This account moreover provides a plausible justification for the interventions made in the following two cases.

In *A Local Authority v A & Anor* [2010][90] an individual was found to lack capacity to make decisions regarding contraceptive treatment because the relational dynamic between her and her husband was thought to impede her ability to 'use and weigh'. Mrs A was a 29-year-old woman with a learning disability and significant cognitive impairments. Prior to her relationship with Mr A, the evidence pointed to a relatively

[90] *A Local Authority v A & Anor* [2010] EWHC 1549 (Fam).

well-adjusted, sociable woman who maintained good relationships with her family and care support workers from the Social Services. A key reason for her current level of functioning was the fact that she had been raised in a loving, supportive environment. Two pregnancies had resulted from previous relationships – both children were taken at birth due to her incapacity to care for them, and she subsequently received a monthly contraceptive injection. Mrs A had been attending college at the time when she met her husband, which she enjoyed because it 'helped her learn things' and she could see her friends.[91] She also did voluntary work in a daycare centre for children.

Following her marriage to Mr A, various expert testimony expressed concerns about his 'controlling behaviour in respect to Mrs A' and that she was being 'de-skilled' by her husband.[92] How Mrs A's self-understanding changed in light of her relationship with Mr A is particularly instructive. She described herself as unhappy at home, she frequently cried at college, and appeared 'defensive [and] feel[ing] quite helpless regarding the situation she is currently in'.[93] Mr A prevented visits from social workers, and Mrs A also stopped attending college as a result of his objections. Mrs A's social worker stated, 'when Mrs A [was] asked a question, she often look[ed] at Mr A to see what his response [was] and that sometimes she [would] not answer at all until he ha[d]. . . . Her experience [was] that Mrs A tend[ed] to go along with whatever Mr A [said]'.[94] Mrs A's choice surrounding contraceptive use reflected this relational dynamic. Her refusal of the injection was primarily down to fears about her relationship with Mr A. She stated that '[Mr A] will not let me take [contraception]. If I do not have a baby, he will kick me out and I will be homeless.'[95] At one point when she thought she was pregnant, she was described as 'really scared' where she stated several times to the course coordinator at her college 'she did not want a baby but that he did and the only way to keep the relationship was for her to have a baby'.[96]

Adherence to the procedural model in the functional test would suggest that Mrs A did in fact have capacity to decide about contraception. Two expert witnesses argued precisely along these lines. Despite Mr A's influence on the decision-making process, Mrs A had understood all the elements needed for a finding of capacity, such as 'prognosis, diagnosis and an understanding of the medical treatments involved, the common methods, how they are used and the possible side-effects'. Indeed, Mrs A

[91] Ibid., paras. 29, 38. [92] Para. 17. [93] Para. 21.
[94] Para. 34. [95] Para. 37. [96] Para. 18.

'volunteered knowledge of the pill and the coil', which indicated that 'she could understand and weigh-up the side effects'.[97]

Rather strikingly, however, Mrs A was found to lack capacity despite the fact that her medical understanding about the issues surrounding contraception satisfied the conditions of procedural reasoning. Disagreement emerged as to whether capacity required a social understanding of the ramifications of childbirth and issues surrounding parenthood in addition to the medical understanding, though this was not the deciding issue. Instead, it was concluded that Mrs A's refusal of contraception resulted from so much relational pressure that her capacity to decide was impaired. Bodey J stated that 'the completely unequal dynamic in the relationship' suggested that 'her decision not to continue taking contraception is not the product of her own free will' and 'she is unable to weigh up the pros and cons of contraception because of the coercive pressure under which she has been placed both intentionally and unconsciously by Mr A'.[98] Evidence supporting the judge's finding of incapacity included 'Mrs A's dependence on [Mr A] and fear of rejection; her suggestibility and wish to please him; his lack of insight to the true extent of her difficulties', as well as Mr A's 'rigid views about their status as husband and wife; his own wish to start a family, which is to be fully respected; and the fact that he has never yet felt included in the decision'.[99]

Similarly, in the case *Re MP; LBH v GP* [2009][100] a 29-year-old man with a minor learning impairment was found to lack capacity to decide living arrangements and familial contact as a result of his relational context. However, his marked improvement since he had been removed from his mother meant his views had considerable significance in the determination of best interests. During the period MP lived with his mother, he had endured an appalling level of neglect: he had been left alone in a locked room with a bucket in which to urinate and defecate whilst his mother was at work. He and his mother lived in a hostel for the homeless following their eviction in the previous flat, and he was told not to leave the room in case their belongings would be stolen: 'MP [was] left alone all day. He [sat] behind the bedroom door on a small stool, some times half-naked . . . in a hunched position much of the day, only moving or stretching when he [went] to the bathroom, such

[97] Para. 45. [98] Para. 72. [99] Para. 73.

[100] *Re MP; LBH v GP* [2009] Claim No: FD08P01058; thanks to Alex Ruck Keene for drawing my attention to this case.

that his feet ache[d] and appear[ed] swollen on occasion.' As sustenance for the day, his mother left a small pot of tea for him as breakfast and lunch.[101]

Like the assessment of Mrs A, the point of entry for medico-juridical evaluations of MP's relational environment was not only the obvious physical and material neglect, but the emotional and psychological mistreatment that meant he was unable to express, let alone critically reflect on, his self-referential responses. MP was entirely dependent on his mother, 'whose attitudes and beliefs influence[d] his ability to make decisions' and he 'appear[ed] frightened and [would] not talk or answer questions when is mother [was] present, or, at most, [would] only say what he [was] told to say by [his mother]'.[102] His removal from his mother to social care dramatically enhanced his decisional capacities, whereby the exercise of his self-understanding and articulatory powers became much more apparent. MP had 'indicated that he would wish to go out into the community to make friends and that he would like to meet people of his own age'.[103] In addition, 'he express[ed] repeatedly a wish to remain living there and not return to live with his mother. His physical and psychological health ha[d] improved markedly since he ha[d] been there. His attitude to contact at present [was] at best ambivalent and at worst hostile. He often [said] . . . that he [did] not wish to see his mother anymore at all'.[104] He further expressed hopes that he would have his own home, a job, and a girlfriend.[105]

These cases are instructive for my purposes for three reasons. First, the judgments posit a link between mental capacity and relational pressure or mistreatment. The Mrs A case is particularly striking in this respect. A finding of incapacity seems incorrect according to conventional legal interpretations of the MCA, particularly given increasing emphasis on the causative nexus between mental impairment and the inability to decide since *PC and NC v City of York Council* [2013].[106] However, the supposed invalidity of the Mrs A judgment is questionable when subject to cross-jurisdictional comparison, most notably its interpretation in the *Re BKR* decision within the context of Singapore's MCA.[107] There it was asserted (rightly, in my view) that determinations of capacity must consider 'actual

[101] *Re MP; LBH v GP* [2009]. Para. 7 (4.1). [102] Para. 7 (4.9). [103] Para. 7 (4.6)
[104] Para. 17. [105] Para. 28.4. [106] EWCA Civ 478.
[107] *Re BKR* [2015] SGCA 26. The comparison is interesting because the MCA in Singapore effectively mirrors the MCA 2005 of England and Wales. The decision articulates insightful ways of interpreting the 'causative nexus', as well as the interaction between mental capacity and external pressure/influence.

circumstances' – namely the interaction between external pressure and mental impairment – rather than a 'theoretical analysis assuming P's emancipation from all external pressure and influence'.[108] Whilst it was submitted that subsequent decisions in England and Wales[109] had re-classified the Mrs A case under inherent jurisdiction rather than the MCA, the Singapore Court of Appeal concluded that 'there was no re-classification of *Re A*, still less a disapproval of it'.[110] The assertion of the court's inherent jurisdiction 'cannot mean that *Re A* is accordingly to be treated as a decision that may be defensible as an exercise of the court's inherent jurisdiction but not of its jurisdiction under the relevant statute' nor suggests that 'Bodey J in *Re A* ought not to have come to the conclusion that Mrs A lacked capacity or that his analysis was otherwise erroneous'.[111]

Second, both judgments illustrate how medico-legal evaluations of the personal-supportive space of reasons (i.e. Bodey J's analysis of the relational dynamic between Mr and Mrs A; Coleridge J's assessment of MP's treatment by his mother) can hinge on the individual's expressed understanding of her- or himself. Mrs A's and MP's way of seeing themselves, the narratives they use to express their values and perspectives, ultimately function as a window into the relational dynamic between them and their support network. This insight has evaluative implications: for instance, that MP's and Mrs A's self-expression was impeded out of fear of their primary carers would be an alerting feature for medico-juridical assessors examining the relational context. The fact that an assessor is 'alerted' to something further presupposes the existence of, and implicit appeal to, background normative values about the role that relationships have in developing or impeding individuals' autonomy. Already, medico-legal evaluations of the MP's and Mrs A's relational context are neither purely procedural nor value-neutral.

This leads to my third point: the medico-legal assessment of the Mrs A's or MP's relational context is only possible when the articulatory-disclosive mode of reasoning is drawn upon. The overlap in both judgments is revealing in this respect. Both cases focus on divergent parts of the statute – one on capacity, the other on deciding best interests – which would initially suggest different models of reasoning ultimately ground their respective judgments. That a best interest decision instinctively

[108] Ibid., para. 98. [109] *Re L (Vulnerable Adults: Court's Jurisdiction)* [2014] Fam 1.
[110] *Re BKR* [2015], para. 93. [111] Ibid., para. 94.

utilises this more substantive mode of reasoning I have been forwarding is hardly new or surprising. The MCA stipulates that a best interests judgment must take into consideration a person's past and present wishes, feelings, beliefs, and values, as well as engage with the relational context of the individual (those who care or are interested in the individual's welfare). In other words, the deliberation of the substitute decision-maker must acknowledge the relational and value-laden character of the frameworks situating the individual, as well as how these frameworks impact on an individual's well-being or the development of future decisional capacity. Yet the judgment in the Mrs A case recognises that the same relational and value frameworks must be taken into consideration when assessing decisional capacity, particularly as it impacts on the individual's ability to understand herself and make decisions in light of that self-understanding.

An immediate worry might be that the overlap in reasoning between the two cases reveals that the type of value-laden paternalism characteristic of best interests decision-making has crept surreptitiously into capacity assessment. From this perspective, an appeal to certain normative values about the relational context would be more acceptable in the case of MP as opposed to Mrs A. Yet I don't think this is the case: the fact that there are values being articulated and expressed in capacity assessment is not itself a problem, but it does mean we need to be attuned to and critically reflexive about *which values* are at stake. Recognising the necessity of the articulatory-disclosive mode of reasoning in practical applications of the MCA helps shift attention towards the specific values being affirmed in the primary-supportive space of reasons. These judgments are arguably non-paternalistic due to the value that they affirm: they do not impose an overtly thick conception of the good life but express a commitment to cultivating the autonomy competencies I defended in the previous chapter. Recall that autonomy competencies are about whether an individual has the ability to respond to discrepancies between one's understanding of oneself and the values of one's relational and social environment. It is about sensitivity and attunement to the core values one has, the confidence and self-esteem needed to stand behind or revise them, and requires one to be situated in a social environment that permits and encourages these reflections and revisions. These autonomy competencies, rather than imposing thick goods and values onto the individual, describe the necessary preconditions of the ability to make practical decisions in line with one's core self-understanding.

The decision-making of both Mrs A and MP (at least prior to his removal from his mother) failed to demonstrate these qualities. It is striking, for example, that V in *V v R* and Mrs A were both described as suggestible – yet V's family sought to cultivate the type of self-reflection that is necessary to develop autonomy competencies, so much so that her mother was worried V was overly influenced by her well-meaning advice. Compare this to Mrs A. In contrast to Mr A's wishes, she did not want to have a child; yet dialogue with supportive others was deemed impermissible. Her increased isolation, confinement, uncertainty, and distress were indicators of how the relational dynamic obstructed the development and exercise of these autonomy competencies. Bodey J's recommendations against a court injunction expressed a commitment to promoting a supportive relational environment where the exercise of these competencies would be possible. 'Ability-appropriate' discussion and therapeutic input was to be provided for Mr and Mrs A, taking into account 'Mr A expressed his willingness in the witness-box to allow Mrs A to have free contact with those professionals who have the skills to advise her in an ability-appropriate way about contraceptive issues, provided he is not excluded but involved as well, either concurrently or separately'.[112] Intervention in MP's case was less about providing open discursive space for the individual to explore one's conflicted commitments without fear of reprisal (as with Mrs A) and more to do with providing an environment where basic physical, material, and psychological care needs could be met so that autonomy skills could begin to take shape. His extremely restricted existence with his mother meant that his ability to express and interpret, let alone critically examine, his own values and self-referential responses was virtually non-existent. Ultimately, these cases articulate and affirm the view that autonomy competencies cannot be promoted unless an individual is situated within a relational space which promotes authentic self-reading and self-expression. More specifically, one's embeddedness within a particular space of reasons is unavoidable – but accessibility and adaptability to *alternative* spaces is necessary, either to reconfirm one's values and commitments or to change them as a result of one's critical engagement.

Conclusion

In this chapter I have presented an internal and external critique of the procedural model of rationality in the context of capacity adjudications.

[112] Para. 80.

In particular, the individualism and value-neutrality implicit in this model are deeply problematic. Once we accept that the space of reasons is unavoidably relational and value-laden, the procedural model's focus on formal reasoning, though initially attractive, is both practically unrealistic and theoretically unconvincing. Capacity adjudications are increasingly required to determine how relationships impact positively or negatively on an individual's ability to make decisions. At its core, these types of cases assess whether an individual's reasoning – the ability to understand and express one's reasons to others – reflects the possession of key autonomy competencies, which presupposes a certain kind of relational and value framework. The content of this framework is still relatively vague from my discussion here, though I go in more detail in the remainder of the book. The important point is that the procedural model is unable to help us with this more evaluative task: as a model of reasoning, it fails to consider a patient's reasons and decisions in terms of her expressed self-understanding; it also fails to situate this self-understanding within the relational and value framework. The relational constituents to decision-making are completely occluded. None of what I am saying negates the reality that the spheres in which these judgments about capacity occur are themselves rooted in implicit normative commitments. But this should signal that our efforts should be directed towards achieving greater articulacy and critical transparency about these commitments and evaluative judgements amongst assessors.

5

Ethical Duties of Support and Intervention

In Chapter 2 I argued against the 'will and preferences' paradigm of Article 12 of the Convention for the Rights of Persons with Disabilities (CRPD), and suggested why the external challenge against mental capacity cannot address coherently the safeguarding issues related to relational abuse and manipulation. By contrast, a relational analysis of rights fundamentally questions liberal assumptions about the private and public divide. The justificatory grounding for duties of intervention has been left open-ended thus far, other than the stipulation in Chapters 3 and 4 that third-party assessments of potentially incapacitating relationships must take as its point of departure individuals' articulated self-reading and self-understanding and whether these reflect sufficiently supportive relational contexts that encourage their perceptual and autonomy skills.

I explore the following question in this chapter: in light of a relational concept of mental capacity, on what grounds are third-party interventions justifiable in consented to but disabling relationships involving individuals with impairments? Consider the following case: WMA was a 25-year-old man diagnosed with atypical autism and a pervasive development disorder. He lived with his mother, who had sight and mobility problems. Support workers were concerned that his mother's care was impeding WMA's long-term development and suggested that a move to supported living accommodation would be in his best interests, despite his strongly expressed wishes to remain with his mother. The judge approved of the local authority's intervention on grounds that under his mother's care, WMA had failed to develop core life and social skills, and it was apparent that her low expectations held him back.[1]

Similar cases arise frequently in medico-juridical practice, where decisions sanction third-party interventions into relational environments that are judged as inhibiting the care-recipient's potential for decisional

[1] *A Local Authority v WMA & Ors* [2013] EWHC 2580 (COP), para. 130.

capacity and autonomous agency. These interventions can be construed in different ways. Through one lens, it represents the bald discriminatory (mis)use of juridical power: capacitous individuals are free to make unwise choices about personally harmful, oppressive relationships – why should it be any different for individuals with impairments, especially if such outside involvement goes against their expressed wishes? As we saw in Chapter 2, the 'will and preferences' paradigm of the CRPD makes an argument precisely along these lines, rejecting non-consensual interventions as inherently paternalistic, disrespectful of the autonomy of individuals with impairments. From another direction, jurists such as Jonathan Herring argue that interventions can be sanctioned and indeed strongly encouraged under a framework of care – if we are serious about promoting caring relationships, then protectionist measures must be robust accordingly.

However, neither of these approaches are satisfactory: as we already saw in Chapter 2, the 'will and preferences' framework is too lax, taking as its starting point questionable assumptions about the private/public dichotomy and the primacy of subjective preferences. As we will see here, these issues are endemic to the social model of disability, which is thought to ground the 'will and preferences' paradigm. Meanwhile, Herring correctly identifies the need to break down questionable assumptions about the protected private sphere and individualistic rights, but, as I explain later, he veers too far in the opposite direction, where paternalistic interests about care and relationality can override the individual's expressions of autonomous choice.

In this chapter I want to make a case for a more balanced approach between these extremes: interventions are justifiable in relational contexts where positive interpersonal duties – such as the duty to develop and promote an individual's autonomy skills – are badly neglected. The previous analysis of individual rights and autonomy has filled out core facets of a more relational concept of capacity; the promotion of autonomy competencies within individuals, moreover, should be a crucial value that foregrounds capacity adjudications. Yet a more robust justification for third-party interventions is required in cases like WMA, where individuals with borderline capacity endorse a disabling relationship. We *should* be worried about the scope of medico-juridical judgments on individuals' chosen relationships. To assuage these concerns, we need to ask, in what sense is it an *ethical obligation* to encourage another's autonomy competencies in relationships? What motivates and justifies this obligation? What are the limits on third-party interventions? Without answering these questions, judgments similar to the WMA case amount to unprincipled, discriminatory meddling. Even more importantly, the limits of such interventions remain dangerously unclear, potentially expanding the boundaries of legally sanctioned paternalistic action into the lives of individuals with impairments.

This chapter offers justification of these positive interpersonal duties, and, in situations of their neglect, articulates normative constraints on third-party interventions. Legal provisions in the supported decision-making paradigm entitle individuals with impairments to enabling forms of support in order to exercise and realise their decisional autonomy. If we probe what these supportive mechanisms entail, we discover these involve not just negative, protective constraints, but robust positive interpersonal duties of assistance, which could very well sanction interventions into disabling relationships. I appeal to the philosophy of Kant to carry out this justificatory task, but not for the most common reasons. Kant's ideas are typically heralded as inspiration for contemporary human rights regimes in the popular imagination. As Catherine Dupré sums up:

> The mainstream legal definition of human dignity still draws on the 18th century philosophy of Enlightenment and in particular on Kant's work. . . . At the core of this definition is the concept of autonomy, an exclusively human attribute which distinguishes humans from other beings and allows them to be active participants in the making of the laws which bind them. This is also commonly expressed as a prescription not to treat a human person as an object, to which judges have referred in interpreting human rights legislation. This approach to human dignity is an essential foundation of our current systems of human rights protection.[2]

The Kantian ideal positing equal bonds of respect grounded on universal human dignity and autonomy has had a profound influence in the medico-juridical context. Yet there remains a question mark as to whether individuals with cognitive impairments are excluded from the Kantian moral community.[3] One must meet a highly demanding standard

[2] Catherine Dupré, 'Unlocking Human Dignity: Towards a Theory for the 21st Century', *European Human Rights Law Review* 2 (2009): 193.

[3] For example, see Kittay, 'At the Margins of Moral Personhood'; Peter Singer, 'Speciesism and Moral Status' in Eva Feder Kittay and Licia Carlson, eds., *Cognitive Disability and Its Challenge to Moral Philosophy* (Chichester: Wiley-Blackwell, 2010), pp. 331–44; Martha C. Nussbaum, *Frontiers of Justice: Disability, Nationality, Species Membership* (Cambridge, MA: Belknap, 2007); also Kong, 'Space between Second-Personal Respect and Rational Care', *Law and Philosophy*.

of moral reasoning to be owed equal respect. Individuals with cognitive disorders whose rationality is permanently impaired may fall short of this threshold, meaning inequality would characterise their relationship with those who care for them. Certain passages of Kant's texts do imply this.[4]

Instead, I want to advance an argument that is more Kantian 'in spirit' rather than to the letter. Anthropological observations of unavoidable human interaction, interpersonal dependency, and rational contingency suggest that all individuals who fall short of full autonomy are included in a moral community, leading to relational duties to encourage another's capacities for autonomous practical agency. Positive and negative third-party duties derive from an inevitable tension between ability and aspiration: on one hand, we aspire towards the ideal of temporally extended autonomous agency, yet our abilities remain imperfect, flawed. We simply cannot reach this aspiration by ourselves, making the assistance of others crucial. An imbalance between ability and aspiration can be indicative of neglect or oppression: individuals can be abused when their limitations are overemphasised, whilst they are likewise neglected if the aspirational dimension of agency is ignored. Justifiable, and indeed, obligatory interventions achieve a suitable equilibrium between recognising limitations to our abilities whilst maintaining the aspirational quality of our agency.

It is this slightly dualistic structure where Kant's ideas have much to offer, particularly compared with conventional sources of justification, such as the social model of disability or care theory, as we see in Sections I and II. Section III explores potential pitfalls in following a Kantian argumentative strategy, but these are mitigated in the exploration of negative and positive duties entailed out of an alternative reading of his Formula of Humanity in Sections IV and V. The legal applicability of this strategy is discussed in Section VI, focusing on the *normative constraints* on third-party interventions and duties of assistance in particular. Finally, Section VII addresses worries about discriminatory paternalistic intrusion into the chosen relationships of individuals with impairments.

[4] For example, Kant implies that we respect the autonomy of individuals with the exception of those with undeveloped rationality: one's *own* concept of happiness cannot be imposed on another out of respect for their humanity, *with the exception* of young children and the insane. Immanuel Kant, *The Metaphysics of Morals*, trans. and ed. Mary Gregor (Cambridge: Cambridge University Press, 1996) 6:454 (Prussian pagination).

I. The Social Model of Disability and Supported Decision-Making

For many, the social model of disability takes into account how relationships have a duty to promote the autonomy of individuals with impairments. Unlike the causal link posited in the medical model, the social model separates impairment (which is a feature of the body and its functioning) from disability (which is socially constructed and the result of surrounding oppressive structural and environmental features). Focus is drawn away from the biological 'defectiveness' of the individual with impairment and towards the social structures and conditions that make impairment 'disabling'. As a result, respect for the autonomy of individuals with impairments involves dismantling environmental barriers that impede their full integration and participation in society.

The practical influence of the social model of disability has been profound. It has been used as a rallying cry for disability advocates, fuelling demands for the equal treatment of individuals with impairments. Most notably, the CRPD formally endorses the social model's premise that disablement has a social rather than biological cause. The Preamble states that disability is 'an evolving concept' that 'results from the interaction between persons with impairments and attitudinal and environmental barriers that hinders their full and effective participation in society on an equal basis with others'.[5] This connection between the CRPD and the social model is important for a number of reasons.[6] First, the CRPD's stance of universal legal capacity accords with the social model's claims that (i) participation in one's social and civic community ought to be universal and equal, and (ii) biological features of impairment should not determine whether an individual is warranted deliberative respect and protection of their rights. Second, the obligation to make reasonable accommodations for individuals with impairments under the CRPD seems premised on the social model's claim that widespread institutional and structural change is required for the realisation of rights.

Supported decision-making is one form of 'reasonable accommodation' that has been much discussed. As I explored in Chapter 2, Article 12

[5] *CRPD*, Preamble (e).

[6] See R. Kayess and P. French, 'Out of Darkness into Light? Introducing the Convention on the Rights of Persons with Disabilities', *Human Rights Law Review* 8:1 (2008): pp. 1–34; Paul David Harpur, 'Embracing the New Disability Rights Paradigm: The Importance of the Convention on the Rights of Persons with Disabilities', *Disability and Society* 27:1 (2012): 1–14.

states the obligation to provide 'all measures to the exercise of legal capacity', expressed through 'respect[ing] the rights, will and preferences of the person . . . free of conflict of interest and undue influence', of which are 'proportional and tailored to the person's circumstances'.[7] Decisional support under Article 12 can be interpreted in a minimal and maximal sense. The minimal sense could simply be respecting an individual's expressed will and preferences, meaning Article 12 effectively functions as a buttress against unwarranted third-party intrusions and thereby protects an individual's right to autonomy. We saw earlier how the 'will and preferences' interpretation eventually retreats to this position. The minimal interpretation is not the most natural reading, however. As Chapter 2 discussed, the provision of decisional support appears amenable to a maximal reading, comprising a combination of macro- and micro-duties, ranging from large scale policies, legislation, and state duties, to relational duties of advocacy, provision of informal networks, and interpersonal support.[8] Probing deeper, the fulfilment of these obligations seems to presuppose a *prior* interpersonal duty to help develop an individual's autonomy competencies – for without the socially acquired confidence, self-knowledge, and self-esteem, the ability to make authentic decisions that accord with one's values remains compromised.[9]

The social model's starting premises purportedly lead to both macro-duties and certain micro-duties, such as the provision of interpersonal networks of advocacy. In particular, they focus on how the social construction of disability sanctions the unequal treatment of individuals with impairments. Social barriers that create or perpetuate disablement should be actively removed as a result. However, the social model's over-simplification of the impairment/disability dynamic weakens its ability to ground CRPD positive interpersonal obligations to support individuals' autonomy skills. From the social model's perspective, impairment as a value-neutral (neither desirable nor undesirable) trait means social integration of disabled individuals merely requires structural changes, such as more inclusive civil rights, political and economic participation; efforts to alleviate and treat impairments are distrusted or even discouraged accordingly. In Harpur's words,

[7] CRPD Art. 12.4. [8] See Gooding, 'Supported Decision Making', p. 10.

[9] As seen in Art. 24.1(a), (b), (c), which states the importance of including individuals with impairments in education so that their 'human potential and sense of dignity and self-worth', their 'personality, talents, and creativity, as well as their mental and physical abilities' can be developed to their fullest potential.

> The problem with policies guided by the medical model is that such policies place undue attention upon 'fixing persons with disabilities'. Medical model policies often do not recognize that a person with a disability has the capacity to live a fulfilling life with a disability. Such policies continually try to 'improve' a person's physical or mental state rather than focusing on other important public issues such as the removal of environmental barriers in society or providing support to enable the person to exercise other rights.[10]

I am not denying that individuals with impairments can flourish without someone 'fixing them'. However, a Catch-22 situation emerges once obligations of decisional support are grounded in the social model: on one hand, universal legal capacity emphasises protecting the negative liberty and civil rights of persons with impairments, presupposing a type of formal equality based on the essential similarity between individuals. Rights thereby function as a protective injunction against state-sanctioned interventions. This motivates arguments about the 'dignity of risk', whereby third-party interventions in the relationships of 'incapacitous' or 'vulnerable' individuals amount to unequal, discriminatory violations of their right to make risky, personally harmful decisions.

But to properly grasp the normative obligations entailed in the provision of decisional support, differences that are residual from impairment must be taken into account. Tom Shakespeare argues convincingly that disability involves a complex interaction between factors both intrinsic and extrinsic to the individual: residual difficulties caused by bodily impairment can affect the extent of one's disability, even *after* social barriers have been removed, indeed, even within the most accessible society.[11] The social model ignores this crucial point, and this explanatory shortcoming leads to an incomplete account of the normative obligations entailed in supporting another's decisional autonomy. The priority of non-interference that formal equality and civil rights protect is not a given: it will depend partly on the person's impairments and could have an unequal impact accordingly. Equal respect for individuals therefore does not always imply a quietist acceptance of impairments or relegating its harmful effects to matters that should be attended to in the private sphere, particularly if this entails ignoring an individual's potential capacities and abilities should aspects of her impairment remain untreated. Equal respect may well involve improving an individual's physical and mental state.

[10] Harpur, 'Embracing the New Disability Rights Paradigm', pp. 2–3.

[11] Tom Shakespeare, *Disability Right and Wrongs* (London: Routledge, 2006), pp. 54–67.

For example, my beautiful niece, Ava, was diagnosed with Rett syndrome. This disorder involves a range of physical and mental impairments, one of which is difficulty feeding. It seems to me that this physical impairment is not value-neutral in the sense the social model claims – it would be a gross neglect of respect and care if my niece's carers did not bring her to therapy to help her learn to chew and swallow her food. Appropriate support must therefore take into consideration the residual difficulties and undesirability intrinsic to certain impairments that subsequently sanction its treatment or mitigation. Understanding the positive duties underlying relational support often assumes a more holistic understanding of how *both* individual and social factors contribute to disability and social integration. Actively removing social barriers is, at times, only a partial response to disabling factors: the residual effects of impairment can impede the development of key autonomy-enhancing personal traits (such as self-esteem, self-confidence, and personal motivation), meaning that a medicalised approach of treatment and rehabilitation may well be morally appropriate in such circumstances.[12]

Similarly, intervening in an individual's chosen relationships might be unjustifiable and stigmatising in certain cases, but not doing so can amount to another form of avoidable neglect.[13] One might argue that a more sophisticated version of the social model can accommodate this intuition, however. Carol Thomas' modified social relational perspective agrees with the social model that disability should be understood as 'restrictions of activity [that] are socially imposed [and] wholly social in origin',[14] though its primary focus on structural barriers neglects how disability emerges out of relational oppression and the exercise of power for the purpose of exclusion. Beyond structural oppression are relational forms of power that have a psycho-emotional impact on a person's self-conception and thereby contribute to the process of disablement. This modified social model thereby grapples with the subjective experience and relational causes of disablement whilst recognising that restrictions

[12] Ibid., p. 59 and Leontine van de Ven et al., 'It takes two to tango: the integration of people with disabilities into society', *Disability & Society* 20 (2005): 311–29.

[13] Susan Dodds also makes this point in her example of the deinstitutionalisation of people with disabilities in Australia. See Dodds, 'Dependence, Care, and Vulnerability', in Catriona Mackenzie et al., eds., *Vulnerability: New Essays in Ethics and Feminist Philosophy* (Oxford: Oxford University Press, 2014), pp. 199–200.

[14] Carol Thomas, 'Rescuing a Social Relational Understanding of Disability', *Scandinavian Journal of Disability Research* 6 (2004): 29.

of activity can occur as a result of impairment (what Thomas refers to as 'impairment effects').[15]

But whether Thomas' modified theory can provide coherent justification for positive relational duties towards individuals with impairments is doubtful. On one hand, Thomas is completely right to think about how relational power and oppression contribute to the psycho-emotional experience of disablement, as we saw in Chapter 3's discussion of socialising influences on developing autonomy competencies and self-constitution. On the other hand, in making relational influences coextensive with oppression or exclusion, Thomas' ring-fenced definition of disability falls prey to the same problems intrinsic to more structural accounts of the social model. Even if these relationships inhibit one's potential for autonomous agency through their psycho-emotional impact in Thomas' sense, I am sceptical that this signals substantive disagreement with other social model proponents who argue that individuals with impairments cannot be treated differently from their non-disabled counterparts in their freedom to remain in abusive, self-harming relationships without the threat of outside intrusions. According to Thomas, the remedy to relational oppression is resistance,[16] but what this morally requires of others is unclear. On one level, resistance could imply outside involvement in personal relationships reflecting an unequal power dynamic; on another level, treating the individual with impairments differently in this respect could be perceived as mirroring an unequal power dynamic which ultimately perpetuates prejudicial, disablist perspectives. Thomas' social relational account of disability, like the more structural account, therefore fails to (i) provide an account of how the mitigation of disablement through non-oppressive, relational support might require addressing and treating another's impairment, as well as (ii) capture how the aspirational dimension of social integration implies a positive duty to help develop a person's autonomy competencies and potential abilities – which may sanction interventions in self-harming choices to remain in oppressive, limiting, or disabling relationships.

As I discussed in Chapter 2, these two normative intuitions have already been implicitly applied in legal practice both in and outside mental capacity law. Interpretations of ECHR Article 3 and 8 rights frequently rest on the articulation of positive interpersonal duties to help develop an individual's autonomy competencies, whether this means addressing certain disabling features of one's impairment or intervening in a

[15] Ibid. [16] Ibid., p. 32.

disabling relational environment. Munby J (as he was then) in the already mentioned pre-MCA case *Sheffield City Council v S* [2002] justified intervening in a private father-son relationship on grounds that the son's potential for greater autonomy was impeded under his father's care. Benefits such as 'the removal of risk of physical ill-treatment; the enhancement in his emotional well-being; the improvement in his social life' meant that S would have 'increased opportunities . . . not merely to develop his potential but also to increase his social contacts and to engage in a wider range of activities than he would enjoy were he to return home'. Munby argued that Article 8 rights 'impos[e] on the State not merely the duty to abstain from inappropriate interference, but also, in some cases, certain positive duties'.[17] Similarly, the judicial decision to place a 24-year-old man into independent living accommodation contrary to his expressed wishes in *A Primary Care Trust v P & Ors* [2009] rested implicitly on a notion of positive obligations to promote and develop P's autonomy competencies, even if it meant opposing his decision to remain within an oppressive, 'unhealthily enmeshed' relationship.[18]

There is a lot to unpick in these cases – and indeed, it is not immediately obvious that these were the *ethically right* judgments. The important point is, whilst space must be accorded for 'the dignity of risk', blanket prohibitions to third-party interventions in disabling relationships disregard the defensible moral intuitions that often underlie these actions. The nature of impairment can sometimes lead to greater dependency in relationships; this in turn heightens one's situational vulnerability to the relational context, which, at its best, can enhance one's decisional autonomy, or at its worst, undermine one's potential for developing autonomy skills as well as the aspiration of social integration. What follows out of this line of reasoning is a range of negative and positive duties, the latter of which may very well sanction intervention in a consented-to relationship that fundamentally undermines the potential for decisional capacity. Justification for such positive duties has been lacking thus far, particularly if we start from the premises of the social model.

II. Care Theory

Another prospective approach towards the justification of positive duties might be found through the resources of care theory. Jonathan Herring

[17] *Re S, Sheffield City Council v S* [2002] EWHC 2278 (Fam), para. 33.
[18] *A Primary Care Trust v P & Ors* [2009] EW Misc 10 (EWCOP), para. 25.

applies the ethics of care to interpret practices in the law, to restructure our understanding of its function and purpose. I do not want to get into a detailed discussion of care theory, but it is worthwhile explaining the manner in which Herring uses this theory to justify third-party legal interventions in abusive contexts. Herring starts with the premise that all individuals are in relations of dependency and care – but contrary to traditional autonomy perspectives, this dependency in itself points to the good of giving and receiving care, of cultivating the virtue or disposition of care.

For Herring, care has four markers.[19] First, *meeting needs* involves addressing bodily, rational, and emotional needs and capabilities. Crucially, meeting needs through care does not require face-to-face interaction but simply involves 'arranging for another's life'[20]; this requirement likewise emphasises that care is often practical, bodily orientated work. The second condition of *respect* surrounds an attitude where the humanity of another is recognised so that the particular needs and wishes of another are attended to through dignified treatment. Third, *responsibility* involves accepting a 'willingness' or 'responsibility' to offer care for a person, it is about the 'appropriate sharing of the burdens of care', particularly given that individuals' capacity to offer care is often finite, focused on a limited circle of others.[21] Finally, *relationality* posits that care is reciprocal, breaking down the barrier between 'carers' and 'cared for'. This allows both parties in a caring relationship to be empowered and enabling; it likewise stresses the interdependent, mutually vulnerable quality of relationships, even when one individual might technically be more 'dependent' than the other. But such dynamics can change, with each party taking turns giving and taking at different times.

These markers of care help refocus legal tools towards assessing how decisions affect caring relationships. The law hones in on questions about 'obligations within the context of relationship' rather than the sphere of one's right to do X; it is to make responsibilities in relationship *prior* to the protection and exercise of individual freedoms. In Herring's words, '[o]ur lives are not marked by freedom, but by our responsibilities to others.... It is in our responsibilities that relationships flourish and in our relationships that we flourish.'[22] Because the law can effectively promote caring relationships, it will therefore have particular obligations to enable or discourage certain relationships, or to help encourage people

[19] Herring, *Caring and the Law*, pp. 17–22. [20] Ibid., p. 17.
[21] Ibid., p. 20. [22] Ibid., p. 62.

to discover alternative caring relationships. If an ethics of care is indeed the grounding virtue for exercising and promoting certain legal rights, then it does seem possible to derive certain positive duties of intervention, particular in cases where individuals are in explicitly uncaring, abusive relational contexts.

I am broadly sympathetic with Herring's ambitions and share some of the same views: for example, the importance of relationality as a lens through which we both conceptualise personal autonomy as well as the normative function of legal rights. I likewise stress the responsibility of surrounding others (as the next chapter shows) and share concerns about the mistreatment of individuals with impairments, particularly in circumstances where dynamics of power can be used to disable and manipulate. But I remain sceptical that care as a concept can support the weight of Herring's claims, or indeed justify grounds for third-party interventions in abusive contexts involving individuals with impairments.

In the first instance, care appears to be a 'catch-all' concept that renders itself meaningless: it is meant to encompass an orientation of emotional and dispositional concern for others and relationships, whilst seeking to protect and respect the individuality and autonomy of persons; it is meant to capture dependency and varying levels of interdependence, yet likewise posits equality and reciprocity. Each of Herring's four markers of care can lead to incommensurable responses to relational abuse: the welfarist concerns of meeting needs could conflict with the second condition of respect. How does overriding an individual's wishes to remain in an abusive relationship reflect an orientation of respect for the subjective claims of the person? Likewise, responsibility does not always cohere with relationality: recognising the burden of care can often compromise the mutuality and reciprocity of a relationship.[23] Given the often incommensurable moral orientations or outcomes each of these concepts might recommend, it seems to me that it is *these concepts themselves* rather than care that shoulders the justificatory burden accordingly. And of course, Herring's markers might be commensurable in certain scenarios – but again, that might result from more favourable circumstances rather than prove the breadth and coherence of care as an ethical theory.

My second concern about care theory is more pressing. Similar to the social model's value-neutral view of impairment, care theory at times ignores that dependency or interdependence is not always value-neutral

[23] This makes the dialogical conditions especially important in such circumstances, as I discuss in the next chapter.

or equal; that the level of dependency can lead to intrinsic, undesirable relational inequalities. In fact, Herring goes further than positing the value-neutrality of these concepts, arguing that '[v]ulnerability and dependence are not only inevitable parts of humanity . . . they are to be welcomed. They are virtues, not vices'.[24] However, this seems to be deriving a contestable normative claim from a natural fact about human nature: it is true that these features pertain to all human experience at some point in our lives, but it is not necessarily the case that it is likewise desirable from a normative perspective, that it is a goal that we ought to aim towards. An unavoidable feature of our experience does not by default make it *valuable*. We might question the value of rugged self-reliance and independence, but this need not mean we also ought to judge dependence and vulnerability as intrinsically worthwhile virtues. Herring speaks of the 'merging of identities' that occurs in care, where boundaries between 'me and you' and 'my body and your body' become fluid. Thus, 'talk of one party having power over the other or an unequal contribution makes no sense, in the closest of caring relationships, where the two merge together as one'.[25] But this overstates the case of relationality, particularly in light of my discussion of developing self-constitution in Chapter 3. Herring's version of care theory is thus potentially susceptible to the dangers of appropriation and assimilation of others, particularly in the context of impairment. We need to recognise and maintain the separateness and otherness of persons to truly appreciate their individuality and unique voice. I discuss this in more detail in the next chapter.

From these general objections to Herring's account, I remain sceptical that care as a concept is sufficiently coherent to justify duties of support or intervention, particularly given that third-party interventions are often motivated by a more formal, impersonal sense of duty rather than some kind of caring bond with the victim of abuse. Herring describes the relationality of care as involving aspects such as 'touch, an expression, the slightest smile [which] can convey great warmth'.[26] These aspects are important for the type of engagement in the primary-supportive context. I am not disputing this. However, the mode of engagement for intervention is a decidedly different matter; third-party interventions demand a more impartial, principled grounding so that it applies whether or not the third-party has a caring bond to the person in question. Even where there may be caring relational bonds, it is unclear on this account what intentions or actions are permissible or impermissible between individuals.

[24] Ibid., p. 55. [25] Ibid., p. 25. [26] Ibid.

III. Kantian Autonomy and Dignity

So far I have examined and dismissed the social model and ethics of care as possible grounds through which duties of assistance or intervention may be justified. Both go too far in opposite directions: the social model's account of impairment fails to account for positive duties to intervene in abusive contexts that may override an individual's subjective wishes; conversely, the care perspective may sanction overly paternalistic actions due to its overemphasis on relationality at the expense of the separateness of individuals. Ultimately, justification for duties of support must also appreciate the contextual, unique features of individuals' embodiment and dependency whilst maintaining the normative significance of developing autonomy competencies examined in Chapter 3. Here I propose a Kantian alternative to fill in this gap.

Inspiration for core foundational medico-juridical values, such as dignity and autonomy, is often found in Kant's philosophy. His principle of 'equal concern and equal respect for all human beings',[27] for example, means that individuals are respected as autonomous choosers and should never be treated merely as a means, but as ends in themselves.[28] The resonance of such principles has contributed to a 'revival [and] increasing influence of Kantian thought'.[29] As Baroness Hale explains, mental capacity legislation and human rights charters embed the Kantian moral intuition that '[e]very human being is important for her own sake and not for the sake of what she can do for others. Every human life has an intrinsic value, irrespective of its usefulness to others, or even of its moral worthiness'. Despite differences in rational capacities and abilities, 'the old, the sick and the disabled have value even though they can no longer be of much use to others, if indeed they ever have been.'[30] In other words, dignity as an inalienable attribute of all humans indicates that all individuals are warranted equal concern and respect, regardless of their mental capacity, age, or diagnosis.

Kantian autonomy and dignity are presumed as value-neutral, egalitarian, inclusive concepts; this popular usage may be unwarranted, however.

[27] See Brenda Hale, 'Dignity', *Journal of Social Welfare and Family Law* 31 (2009): 101–8; also Suzy Killmister, 'Dignity – Not Such a Useless Concept', *Journal of Medical Ethics* 36 (2010): 160–1.

[28] Christopher McCrudden, 'Human Dignity and Judicial Interpretation of Human Rights', *European Journal of International Law* 19 (2008): 659.

[29] G. P. Fletcher, 'Human Dignity as a Constitutional Value', *University of Western Ontario Law Review* 22 (1984): 171.

[30] Hale, 'Dignity', p. 104.

For Kant, autonomy expresses the capacity for moral self-legislation: namely the ability to reason and will in accordance with universal principle as articulated in the Categorical Imperative – 'act only on that maxim through which you can at the same time will that it should become a universal law'.[31] This Formula of Universal Law (or CI-test) describes a procedure by which the moral permissibility of one's underlying maxims is tested. A maxim is an underlying principle of willing which manifests itself through particular choices; for example, a maxim to appreciate beauty might mean the decision to go on a nature walk or look at Rembrandt paintings in an art museum. At a basic level, this imposes formal constraints of consistency on one's reasoning so as to ensure *subjective* coherence between one's overarching principles, values, beliefs, and particular desires or situational choices.[32] But at a more important level, it requires *universal consistency*. For example, the maxim not to repay my debts despite promising otherwise might fulfil the conditions of subjective consistency (it accords with my overall self-interest or helps me achieve my end of obtaining funds). Yet such a maxim is fundamentally self-defeating according to the CI-test: once such a maxim is universalised, the very purpose of promising is made obsolete since the promises of another could never be trusted. The universality test therefore functions as a negative constraint against maxims and principles of action that others cannot likewise adopt.

Whether many of us, let alone individuals with impaired rational capacities, can meet the exacting standards of moral self-legislation should already be questionable. Other formulations of the CI-test exacerbate worries about the overly demanding nature of Kantian practical rationality, such as the Formula of Humanity: 'Act in such a way that you always treat humanity, whether in your own person or in the person of any other, never simply as a means, but always at the same time as an end'.[33] Compared with 'things' which only have relative value (in the sense that we can manipulate them for our particular prudential or skilful

[31] Kant, *Groundwork of the Metaphysics of Morals*, trans. H. J. Paton (London: Routledge, 1948), 421[52].

[32] See Onora O'Neill, *Constructions of Reason: Explorations of Kant's Practical Philosophy* (Cambridge: Cambridge University Press, 1990), pp. 81–104. It is debatable, however, whether the normativity of categorical imperatives demands subjective consistency in the same manner as that of hypothetical imperatives. See Camillia Kong, 'The Normative Source of Kantian Hypothetical Imperatives', *International Journal of Philosophical Studies* 20 (2012): 661–90.

[33] Kant, *Groundwork* 429[66–7].

purposes), persons have a nature which 'already marks them out as ends in themselves... and consequently imposes to that extent a limit on all arbitrary treatment of them (and is an object of reverence)'.[34] Contemporary medico-juridical justifications of informed consent and patient autonomy often appeal to the Formula of Humanity, yet it also has the most troubling implications for those with cognitive impairments. 'Persons' or 'ends in themselves' denote rational beings capable of end-setting agency, whose reasons accord with the categorical imperative. Such moral motivation warrants respect for autonomy, confers dignity onto a person, and binds together the moral community of end-setting agents.[35] Because morality is the supreme value, when we value morality above all else and follow the moral law, our moral standing is elevated accordingly. Conversely, the individual who is cognitively impaired, or who can reason prudentially but is incapable of achieving the sort of moral motivation and practical reasoning characteristic of the good will, would be conceptually relegated to the 'margins of personhood'. Those individuals incapable of the necessary deliberative skills to develop their moral agency in the form of self-legislation therefore remain outliers of the moral and deliberative community.[36] The obligations and duties owed to them would be grounded on paternalistic beneficence rather than respect for their autonomy and dignity. In short, this particular Formulation of the CI-test suggests the unequal moral status of individuals who lack the capacity for moral self-legislation. The popular appeal of Kantian dignity to establish the equal worth of all persons, regardless of one's impairments, faces genuine problems once the connection between autonomy, dignity, and moral self-legislation is probed further.[37]

The textual support for this more value-laden, moralistic reading of Kant makes its normative implications deeply unappealing. Kant stresses residual disabling effects of impairments to the point that there appears no space for respecting the autonomy of such persons. The premises of Kantian dignity and autonomy thus seem to entrench discriminatory, disabling attitudes, leading to far more damaging consequences compared with that of either the social model or care theory. These pitfalls emerge,

[34] Ibid., 428[65].

[35] Oliver Sensen, 'Kant's Conception of Human Dignity', *Kant-Studien* 100 (2009): 309–31.

[36] For more on this, see Michael Rosen, *Dignity: Its History and Meaning* (Cambridge, MA: Harvard University Press, 2012). This problem also afflicts contemporary versions of Kantian contractualism: see Kong, 'The Space between Second-Personal Respect'.

[37] See also Richard Dean, *The Value of Humanity in Kant's Moral Theory* (Oxford: Clarendon, 2006).

however, only if the more common Kantian starting points of universal dignity and autonomy are adopted as the basis for what is owed to individuals with impairments. I want to adopt a different strategy that is more Kantian 'in spirit' to sidestep these dangers.

Kant's highly moralised account of practical reasoning is obviously inapplicable as a standard of rationality in the capacity test itself. Yet his argumentative strategy nonetheless provides a fruitful way to understand and justify *normative interpersonal obligations* owed to individuals with impairments from *impartial, third-personal perspectives, when the limits of caring bonds are reached.* Certain actions can be found to be impermissible, whilst motives of duty often take over when caring engagement is restricted. In other words, a Kantian framework captures more impersonal dimensions characteristic of medico-juridical practitioners who are *outside* the intimate relational context to a large extent (occupying what I called the 'second-order' space of reasons in the previous chapter) and further articulates the *normative perspective* they need to adopt to fulfil their obligations. Even more importantly, this framework accounts for how interpersonal duties of assistance form a vital bridge between empirical facts about impaired, embodied rationality and human aspiration towards autonomous agency, thus grounding obligations to cultivate the autonomy competencies discussed in Chapter 3. Problems with the social model's oversimplified accounts of the impairment/disability dynamic can therefore be sidestepped whilst reasserting the normative value of autonomy that tends to be minimised in care theory.

The Formula of Humanity contains two crucial, yet frequently neglected anthropological assumptions about the conditions of human rational agency. I explore these in the following two sections. First, the universality test presupposes the *social and interpersonal space of reasons.* The CI-test recognises the *practical, normative* implications of Chapter 4's observation about the socially constituted nature of our spaces of reasons. One's maxims need to coexist with the willing and ends of others given the fact that we are situated within a common social space. This leads to a *negative duty* to ensure that one's own maxims do not undercut another's agency in various ways. The second presupposition is that *human rational powers are unavoidably contingent and limited*: we are dependent, finite, imperfectly rational beings who require the assistance of others. This key assumption leads to *positive third-personal duties*, such as the obligation to foster and assist another's practical agency, on top of negative duties that constrain action. Starting from these two premises allows us to bypass controversial issues surrounding the status of individuals with cognitive impairments, particularly given that universal rational contingency can be said to characterise all human agency.

IV. Negative Duties of Constraint

The injunction to treat persons as ends in themselves presupposes the recognition that we are part of a social, interacting community of end-setting agents. At one level, this makes a simple spatial point about the empirical reality that persons share a common space of coexistence: that coordination between varied ends and maxims, as well as a sphere of negative liberty and non-interference, are needed in order to exercise one's agency. But at a deeper level, the CI-test in the Formula of Humanity demands not just consistency between different maxims in the social space of reasons, but moral awareness of how one's maxims could undercut another's ability to exercise their agency.[38] Recognising another person as an end means acknowledging that she too is an end-setting agent, and what I can subjectively will is constrained accordingly. For example, maxims of coercion and deception treat an individual as a mere means because it impedes another's ability to set her own ends. Deception pre-empts our ability to respond to the true maxims of another person; conversely coercion pre-empts our ability to respond to our genuine maxims. Both types of maxims treat others as an object to manipulate for my own purposes, to impose my will onto someone for my own purpose. A genuine choice between collaboration, assent or dissent cannot be made in both scenarios.[39]

Legal interventions in a person's private relationship recognise how maxims of coercion and deceit fundamentally undermine individuals' practical agency. Consider the important pre-MCA inherent jurisdiction case, *Re SA* [2005][40]: SA was a 17-year-old young woman who had capacity to consent to marriage but was profoundly deaf and unable to speak. Concerns were raised that SA's parents would take her to Pakistan for a forced marriage. This case is significant for a number of reasons, most notably of which was the procedural extension of the court's protective jurisdiction to 'vulnerable adults' who still technically had the capacity to decide.[41] But its underlying justification reveals the intuitive truth

[38] See Barbara Herman, *The Practice of Moral Judgment* (Cambridge, MA: Harvard University Press, 1996), p. 52.

[39] O'Neill, 'Between Consenting Adults', *Philosophy and Public Affairs* 14 (1985): 262.

[40] *Re SA* [2005] EWHC 2942 (Fam). [41] See Appendix 1.

of Kant's observations about the disabling effects of coercive, deceptive maxims on individuals' practical agency. Munby J (as he was then) argued in *Re SA* that inherent jurisdiction could apply to vulnerable individuals subject to conditions of (i) constraint, (ii) coercion or undue influence, and (iii) other vitiating factors that deprived them of decisional capacity and the ability to make a genuine choice. Coercion in this context is defined as when 'a vulnerable adult's capacity or will to decide has been sapped or overborne by the improper influence of another' where it may be 'very little pressure [in order] to bring about the desired result'.[42] Other vitiating factors include deception and misinformation, used to 'reduce a vulnerable adult's understanding and reasoning powers as to prevent him forming or expressing a real and genuine consent'.[43] Of course, the concept of 'vulnerability' itself may be problematic, and whether this is a useful legal category has already been questioned.[44] Indeed, some might argue (unconvincingly, in my view) that moves in subsequent legal decisions towards an emphasis on the intrapersonal cause of mental incapacity is precisely to avoid procedural bleeding between inherent jurisdiction and the MCA.[45] But for my purposes here, it suffices to note the overlap between Munby J's justification of inherent jurisdiction and Kant's observations about the vulnerability of our agency to undermining coercive and deceptive maxims willed by others.

Our rational, practical agency is situated within a common space of reasons, making us interdependent beings. Maxims of non-beneficence or pure self-sufficiency are impermissible as a result. Even as an individual appears to be flourishing without the assistance of others, such a maxim is fundamentally incoherent, as 'many a situation might arise in which the man needed love and sympathy from others' and 'he would rob himself of all hope of the help he wants for himself'.[46] Moreover, pure self-sufficiency depends on numerous contingencies and resources over which one has no control. Whether we are endowed with the skills, talents, and abilities needed to reject or carry on without the help of others often depends on luck.[47] The universality test draws attention to the fact that, even if one possesses the requisite skills and talents to be self-sufficient, its random natural distribution means not everyone will be so fortunate. A maxim

[42] *Re SA*, para. 78 (ii). [43] Ibid., para. 78 (iii).

[44] Michael Dunn, 'When Are Adult Safeguarding Interventions Justified?' in J. Wallbank and J. Herring, eds., *Vulnerabilities, Care and Family Law* (London: Routledge, 2013), pp. 234–53.

[45] *PC & Anor v City of York Council* [2013] EWCA Civ 478.

[46] Kant, *Groundwork*, 423 [56–7]. [47] Herman, *Practice of Moral Judgment*, p. 53.

of non-beneficence would impede the practical agency of others without the skills of self-sufficiency because the supportive conditions by which their end-setting abilities can be exercised would be removed.

V. Positive Duties

To summarise thus far, empirical observations regarding the social space of reasons lead to negative duties, which simultaneously demand consistency and restrict maxims that vitiate the possibility of another exercising her practical agency, through maxims of coercion, manipulation, or non-beneficence. These negative duties, however, express only part the Formula of Humanity's normative implications. Its necessary counterpart revolves around positive duties of support that result from a core, dualistic tension between *ability* and *aspiration* in what it means to be human. At its core, Kantian humanity describes a being that is capable of setting ends of flourishing as well as acting out of moral duty within empirical conditions. This basic definition of humanity expresses two sides of the same coin: on one side is an aspirational, normative ideal that articulates limiting conditions (the idea of humanity-as-end-setting-being constrains our freedom through the negative duties explained earlier) as well as imposes an objective goal of autonomous, end-setting agency that we must actively promote within ourselves and others.[48]

The flip side, however, is a recognition that humanity-as-an-ideal remains an elusive goal given the empirical realities surrounding human rationality and agency: purely self-reliant, rational introspection might sound good in the abstract, but real human agency is messy, complex, and defined by embodiment and dependency. This dynamic within humanity resonates with issues of impairment and disability: on one hand, the aspiration of social participation and integration revolves around the ideal of autonomous persons setting and realising their own ends. On the other hand, the reality of embodiment, impairment, and dependency imposes very real constraints on the possibility of realising this aspiration *through one's own individual efforts.* This dualism between the *ideal* of autonomous humanity and the *empirical characteristics* of our agency expresses an important gap that can be filled only through obligatory duties of assistance. Hence why moves to reduce this gap leave little space for positive duties of assistance. Either individual abilities must be inflated to the point that mutual assistance is unnecessary to reach this aspiration

[48] Kant, *Groundwork*, 431 [70].

(i.e. the social model's value-neutral definition of impairment), or the aspiration itself becomes lowered to the point that the contingent nature of one's agency does not bear on the goal's achievability (i.e. care theory's misguided idealisation of dependency and vulnerability).

Positive duties are simultaneously ends we ought to adopt, and include both self- and other-regarding dimensions. There is a duty to respect the humanity-as-end-setting-being *within ourselves*, and this includes cultivating natural powers of spirit, mind, and body to ensure that one's natural predispositions and capacities can be used effectively in pursuit of one's ends.[49] Unlike the dominance principle discussed in Chapter 3, this duty does not rest on determinate basic interests or conflate potentialities with talents, gifts, or aptitudes. As I mentioned there, we need to distinguish the latter from those potentialities that are vital to the development of autonomy competencies. This duty seeks to promote potentialities in the second sense, namely the *conditions* of one's own practical agency. I have a duty to myself to ensure that my willing does not bind myself to the point that I can no longer set my own ends.[50] However, we don't always value our temporally extended agency; we often make choices that constrain our future actions. Individuals can choose to neglect their natural abilities or disregard themselves as humanity-as-end-setting-being. Nonetheless, this obligation articulates a core intuition about the shape and form of autonomous agency as an ideal towards diachronic, temporally extended agency. We cannot be *morally blamed* for willing maxims that fundamentally undermine our own future agency, but to call that an appropriate aspiration for autonomy would be counterintuitive.

So the duty to respect the humanity within ourselves can be understood from the first-personal perspective, as from the standpoint of the individual in question. From a third-personal perspective, we cannot blame such individuals for its neglect, even as we might praise those who do the opposite.[51] Crucially, this duty has an important link to other-regarding positive duties of beneficence that require us to adopt the happiness of others as our own end. These are interlocking yet non-interchangeable commitments: the duty to cultivate my natural abilities and moral character does not mean it is a duty to perfect others; conversely the duty to

[49] Kant, *Metaphysics of Morals*, 6:392.

[50] Admittedly, Kant's account of the duty of self-perfection can be at times troubling, particularly considering how his unconditional prohibition against suicide flies in the face of many moral views surrounding euthanasia as a compassionate response to cases of profound human suffering.

[51] This is what Kant means in calling the duty of one's perfection an 'imperfect' duty.

promote the happiness of others does not mean it is a duty to promote *my own* happiness. This suggests an other-regarding, outward-looking orientation to these positive duties. A commitment towards my self-perfection means that I will be better equipped with the capacities needed to support and promote the well-being and ends of others; conversely, it removes those common excuses for not assisting others based on passive acceptance of shortcomings within my character: 'I don't have the patience to deal with individuals with impairments; I just can't communicate with a person with a learning disability; I'm just not a sympathetic type of person so can't comfort someone who is suffering'.[52]

In other words, there is an important perspectival, normative shift that occurs when we focus on positive duties: we might consider the first-personal perspective initially (in terms of the question, 'does this decision help actualise my ends-setting agency?'), but its corresponding positive duty to adopt the ends of others as my own implies an impartial, third-personal perspective, which reveals that the *community around us* has certain normative obligations of beneficence to us, and vice versa. This may appear to merely echo entrenched medico-juridical principles of beneficence. However, Kantian inflections to the duty of beneficence differ from contemporary medico-juridical accounts in an important respect. The latter comes across either negatively – 'do no harm' – or paternalistically – welfarist, substituted decision-making based on what is objectively good for another.[53] Yet both inflections are absent in Kant. To simply refrain from impeding the happiness of others would be an incomplete fulfilment of beneficence: we must be active in seeking 'to further the ends of others. For the ends of a subject who is an end in himself must, if this conception is to have its full effect on me, be also, as far as possible, *my* ends'.[54] Beneficence revolves around assisting others without the expectation of personal gain, creating a 'community of mutual aid' where the conditions of another person's end-setting powers are developed through interpersonal support.[55] The duty of beneficence thus means I act in a way that will foster those autonomy competencies that enable her to actively pursue and implement her plans towards her happiness; it means promoting the temporally extended agency of another

[52] See Marcia Baron and Melissa Seymour Fahmy, 'Beneficence and Other Duties of Love in *The Metaphysics of Morals*', in Thomas E. Hill, Jr., *The Blackwell Guide to Kant's Ethics* (Chichester: Wiley-Blackwell, 2009), p. 227, n. 4.

[53] See for example, Stephen Darwall, *Welfare and Rational Care* (Princeton: Princeton University Press, 2002).

[54] Kant, *Groundwork*, 430 [69].

[55] Herman, *Practice of Moral Judgment*, p. 60.

person, *not* substituting my agency for hers or imposing a certain good onto the individual.[56] Emphasis on promoting the conditions of another's agency in this manner therefore constrains the paternalistic impulse to perfect others. This is partly because it is inherently incoherent to perfect another person who is herself an end-setting agent – one cannot impose on her how she ought to perfect her abilities. It is a 'contradiction for me to make another's *perfection* my end and consider myself under obligation to promote this,' Kant argues. 'For the *perfection* of another human being, as a person, consists just in this: that he *himself* is able to set his end in accordance with his own concepts of duty; and it is self-contradictory to require that I do (make it my duty to do) something that only the other himself can do'.[57]

But even more importantly, it would be arrogant to presume that we can perfect another individual. Positive duties illustrate well how the gap between abilities and aspiration constrains paternalistic, arrogant attitudes. On one hand, the extent to which we can control and actualise our self-perfection is limited; we can never fully know our intentions or maxims.[58] And as we saw in Chapter 3, perceptual, bodily challenges can compromise our ability to cope skilfully with our environment, whilst our relational, social context may or may not help us develop the autonomy skills necessary to choose and live authentically. To paraphrase Kant, we might strive to realise our humanity-as-end-setting-being, but we struggle to reach it by ourselves.[59] Our unavoidable embodiment and rational contingency likewise constrains even our attempts to assist others: we are imperfect beings and our capacities for moral action are prone to disturbance and failure. Our abilities remain vulnerable to interpersonal interaction, rejection, and affirmation. We exaggerate our tenuous capabilities in the belief that we can perfect another individual, reflecting instead an attitude of self-importance that functions as a barrier to improving our abilities whilst contradicting the normative intent behind duties of beneficence. Humility must underline our beneficent acts, likewise those negative duties of respect that proscribe treating others with contempt and arrogance, or engaging in defamation and ridicule.[60] Arrogant condescension fundamentally minimises another individual, resulting in the sort of counterproductive psychological harm that thwarts the development of core personal resources necessary for autonomous

[56] Ibid., p. 70; O'Neill, *Constructions of Reason*, p. 140.
[57] Kant, *Metaphysics of Morals*, 6:386. [58] Ibid.
[59] Ibid. 6:446. [60] Ibid., 6.462–8.

agency. Conversely, positive acts of beneficence, such as generosity, gratitude, sympathy, and compassion, as well as avoiding envious, ungrateful, and malicious propensities, are important ways in which one encourages another's ends of happiness and well-being.[61] These acts assume each has his or her own existing ends and merely describe the appropriate attitudinal stance that assists another in achieving them.

Already it should be clear that although the Formula of Humanity takes a universal form, particularistic judgements assessing individual and relational features are vitally important to fulfil its constituent duties. Failure to treat persons as ends in themselves is often characterised as violating a person's autonomy and freedom to set her own end. But *neglect* can likewise represent a failure to treat a person as an end. We neglect individuals when we withhold encouragement, generosity, compassion – key ways that individual agency is promoted. We likewise neglect individuals when we fail to assist or intervene when the opposite conditions prevail. Our vulnerability varies in degree depending on an individual's specific competencies, needs, and relationships. Appreciating these specificities helps determine when a stance of non-interference can best promote another's end, and conversely, when active assistance, through encouragement, guidance, or even unwanted intervention, is appropriate.[62] Properly sharing another person's end of happiness therefore requires one to make situational judgements to determine violations of beneficent duties within the immediate relational context, as well as carry out corresponding beneficent, third-party duties *outside* such relationships.[63] As Herman states, '[t]he point of morality . . . is to regulate what can go on given the vulnerabilities of persons as agents – vulnerabilities we all share as human rational beings *and* vulnerabilities that are specific to our situations and relationships'.[64]

VI. Application to the Law

The discussion of Kantian positive duties of beneficence thus provides a normative lens for third parties to judge and intervene in primary relationships that disable an individual's potential end-setting agency.[65] This is a particularly illuminating perspective through which to interpret certain influential legal judgments pioneered by Munby J (as he was then).

[61] Ibid., 6: 453–61. [62] O'Neill, 'Between Consenting Adults', p. 265.
[63] Ibid., p. 272. [64] Herman, *Practice of Moral Judgment*, p. 205.
[65] O'Neill, 'Between Consenting Adults', p. 272.

For example, the pre-MCA inherent jurisdiction case *Sheffield City Council v S* is notable in how the judgment focused on providing S opportunities to develop his potential autonomy skills.[66] Whereas S 'appeared passive and under stimulated at home' due to his father's 'emotionally confining and stultifying' care',[67] he was reported as 'becoming more confident, more independent and interacting more with everybody', as well as 'had a greater sense of himself... show[ing] potential for developing skills' following his placement in institutional accommodation.[68]

Likewise in *Re SA*, Munby J argued the family's neglect of their positive duties to SA meant that she 'could not give informed consent unless all her special needs [were] taken into account and she fully comprehend[ed] what is being proposed to her'.[69] SA's main source of communication was British sign language, which her parents were unable to use, meaning neither could communicate his or her respective wishes and plans; moreover, these communicative barriers in a foreign environment could potentially undermine her mental health, causing psychological difficulties and extreme distress.[70] The family's actions effectively disabled rather than enabled SA's practical agency, and her unique communicative and autonomy skills were disregarded in light of what her family thought to be best for SA's happiness. Yet SA made clear her values and wishes regarding marriage to officials.[71] Munby J's justification for exercising inherent jurisdiction has strong Kantian echoes. The court had a 'positive duty to assist SA to enter into what will for her be the "right" marriage, with someone who will confirm to her, in a way she can understand, that he understands and agrees with what she wants'.[72] SA's ECHR Article 8 rights sanctioned court intervention, but Munby J argued that this

> stands merely as a *means to an ultimately more important end*. In the final analysis, my concern must be to *enable* this vulnerable young woman to exercise her right to self-determination, specifically her right to marry as enshrined in Art 12 of the [ECHR]. [...] [T]he court has a positive duty to assist SA to enter into what will for her be the 'right' marriage, with someone who will confirm to her, in a way she can understand, that he understands and agrees with what she wants. My ambition is that SA

[66] *Sheffield City Council v S*, para. 118. [67] Ibid., para. 110, 120.
[68] Ibid., para. 110. [69] *Re SA*, para. 30. [70] Para. 15.
[71] She did wish to marry, but not yet; when she did, she anticipated it would be to a Pakistani Muslim man chosen by her parents but approved by her; and she expected her husband to immigrate to Britain as she expressed strong wishes to remain in the same city as her family. Ibid., para. 24.
[72] Ibid., paras. 30–1.

> should lead a happy and fulfilled life as a woman, wife and mother, and it is with the aim of facilitating and achieving this ambition that I exercise my powers under the court's inherent jurisdiction. By taking this course, far from depriving SA of her right to make decisions, I am ensuring, as best I can, that she has the best possible chance of future happiness. I am taking these steps to protect, support and enhance SA's capacity to control her own life and destiny in the way she would wish.[73]

Munby J's statement here reiterates various insights of Kant's Formula of Humanity: human dependency – particularly in light of certain impairments – means that duties of non-interference are only part of our interpersonal obligations. Adopting another individual's end of happiness as my own requires one to promote the conditions by which an individual can exercise her autonomy competencies to realise her own ends. Munby J's important rulings here illustrate that the courts can also adopt this moral stance towards the individual and intervene in the absence of appropriate relational support and disabling decisional conditions, where one's relationships impose heteronomous, inauthentic ends of happiness onto oneself.

These positive duties of beneficent support are not confined just to protectionist interventions, nor do they emerge solely in pre-MCA, inherent jurisdiction cases. Just as the statute came into effect, Munby J decided a case *A Local Authority X v MM & Anor* [2007][74] concerning the capacity of MM, a woman with paranoid schizophrenia, to decide her living arrangements and contact with her partner, KM, of fifteen years. The local authority raised concerns about the latter's power, influence, and potential manipulation over MM, and subsequently claimed that MM should be in supported living accommodation where all unsupervised contact with her partner was impermissible. Munby J condemned such protective measures as disproportionate, unjustifiable, 'a revealing example of the dangers of putting a vulnerable adult's supposed safety before her happiness'.[75] He stated:

> Physical health and safety can sometimes be bought at too high a price in happiness and emotional welfare. The emphasis must be on sensible risk appraisal, not striving to avoid all risk, whatever the price, but instead seeking a proper balance and being willing to tolerate manageable or acceptable risks as the price appropriately to be paid in order to achieve

[73] Ibid., paras. 130–2, emphases added.

[74] *A Local Authority X v MM & Anor* [2007] EWHC 2003 (Fam).

[75] Ibid., para. 156.

> some other good – in particular to achieve the vital good of the elderly or vulnerable person's *happiness*. What good is it making someone safer if it merely makes them miserable?[76]

The judgment criticised the local authority's disregard for the long-standing sexual relationship between MM and KM. The local authority had imposed on MM a situation that could violate her ECHR Article 8 rights, meaning it had a positive obligation to enable MM to continue her sexual relationship with her partner 'in an appropriate and dignified way', with access to appropriate facilities.[77]

Another legal judgment appeals to beneficence in similar ways to those I have explored thus far and shows how it can apply not just to questions of capacity. *Westminster City Council v Manuela Sykes* [2014][78] concerned a best interests decision regarding a dispute about Sykes' care and residence. She was noted to be 'by nature a fighter, a campaigner, a person of passion' who participated in social and political causes, including the rights of dementia sufferers, following her own diagnosis.[79] She had made a living will, which articulated in general terms that 'she prioritis[ed] quality of life over prolongation of life'.[80] Her health and living conditions had deteriorated, exacerbated by the fact that she had become increasingly aggressive towards visiting carers, due to the progression of her dementia. She was admitted to hospital under S.2 of the Mental Health Act 1983 and subsequently discharged to a nursing home rather than her own home, which she vigorously opposed. Eldergill DJ carefully considered Sykes' views and values, having met with her in person at the nursing home. The result is a highly sensitive best interests judgment, which alongside other relevant factors of the law, drew on the principle of beneficence explicitly. Eldergill DJ argued that S.4(4), (6) recognised the fundamental importance of liberty and autonomy even to incapacitated individuals who maintain clear wishes and preferences about where and how they live. He stated:

> Society is made up of individuals, and each individual wills certain ends for themselves and their loved ones, and not others, and has distinctive feelings, personal goals, traits, habits, and experiences. Because this is so, most individuals wish to determine and develop their own interests and course in life, and their happiness often depends on this.[81]

[76] Ibid., para. 120. [77] Ibid., para. 162.
[78] *Westminster City Council v Manuela Sykes* [2014] EWHC B9 (COP).
[79] §5. [80] §5. [81] §10.

Eldergill DJ continued that the law demanded 'objective analysis of a subject not an object' which meant, 'it is *her* welfare in the context of *her* wishes, feelings, beliefs and values that is important. This is the principle of beneficence which asserts an obligation to help others further *their* important and legitimate interests. In this important sense, the judge no less than the local authority is her servant, not her master.'[82] The decision to implement a trial period at home recognised that Sykes may suffer some distress, but the overriding goal was 'to strive for as long as possible to achieve those ends which have given her life its value and which she still wishes to pursue'.[83]

To be clear, I am not claiming that the judgments in these cases themselves invoke a Kantian line of reasoning as justification for intervening in potentially disabling relationships or imposing a particular decision about their care or treatment onto them. But the resonance in language is striking, and adopting a Kantian lens makes the concepts and reasons behind these judicial decisions intuitively plausible, in one sense particularly. Both Munby J and Eldergill DJ's judgments here strike a crucial balance between the extremes of neglect and paternalism. Kant's account of positive duties of beneficence explains precisely why this balance is so important. When we fulfil positive duties of beneficence in the Kantian sense, we do not necessarily engage in paternalistic treatment in a strong sense – rather, it is to treat someone as a potential end-setter, to effectively 'support his status as a pursuer of ends, so that I am prepared to do what is necessary to help him maintain that status'.[84] Paternalistic overtones tend to dominate traditional conceptions of beneficence, where we pre-empt another's projects and ends to impose an objective good. A dichotomy between respect for autonomy and paternalistic beneficence is therefore posited. But as these cases help express, this strict dichotomy doesn't apply all the time. In such scenarios, we need positive duties of beneficence that articulate clear constraints in terms of what one can justifiably impose on an individual in order to promote her happiness. So in the case of SA, the court could not choose a partner for her but justifiably intervened to facilitate the conditions in which she could communicate and implement

[82] Ibid. This reflects Eldergill DJ's approach to judicial decision-making more generally. He stresses the importance of compassion in the law and states that 'the principle of beneficence . . . asserts an obligation to help others further their important and legitimate interests, not one's own'. See Anselm Eldergill, 'Compassion and the Law: A Judicial Perspective', *Elder Law Journal* [2015]: 278.

[83] *Westminster City Council v Manuela Sykes* [2014], §10.

[84] Herman, *Practice of Moral Judgment*, p. 70.

her own plans about marriage; similarly, the local authority's protective treatment of MM clearly went too far in its failure to provide the necessary conditions in which she could realise her own end of continuing a long-standing sexual relationship. Likewise, the temptation to minimise risks or prevent harm to Sykes meant disregarding the values, feelings, and ends she held dear. Decisions in these cases were grounded on positive duties of enabling another's autonomy skills to pursue their own authentic ends through the provision of support and encouragement – duties that the individual's surrounding relationships may have previously neglected.

In sum, my reading of Kant's Formula of Humanity provides a useful justificatory strategy for positive duties of assistance and support. This Formula articulates the normative principles underlying a community of mutual aid, based on inescapable facts about our universal dependency. The gap between aspiration and ability generates interpersonal duties of assistance and support: on one hand is our inherent vulnerability, on the other are aspirations towards the exercise of one's end-setting agency and its constituent autonomy competencies. Our common rational contingency and vulnerable embodiment, combined with our common aspiration towards certain ends, will lead to obligations of relational support to minimise this gap. Such support will involve both positive and negative duties in the form of encouragement, generosity, compassion, and respect in the form of non-interference. By the same token, these duties may well sanction third-party interference, particularly if such means enable an individual to develop crucial autonomy competencies needed to realise their varied ends.

Medico-juridical practitioners are often *external* to the individual's primary relationships. As a result, duties of intervention and support must be able to incorporate and account for the normative perspective of third-party judgement, guided by humility and impartial bonds of duty rather than intimate care. This ability to account for the more impartial perspective is the unique strength of Kant's philosophy. As such, then, I am in no way arguing that his demanding account of moral self-legislation should be included in our concept of capacity: individuals' failure to respect their own end-setting humanity – their temporally extended agency, in other words – may flag up deficiencies in autonomy competencies, but *should not determine capacity in itself*. None of what I am saying displaces the claims in Chapters 3 and 4 that an individual's autonomy competencies, distilled through their self-conception and self-reading, must be the point of departure for all third-party judgements about capacity and relational support. Importantly, recognising human agents as vulnerable, rationally contingent beings helps lessen the conceptual divide between those who have impairments and those who do not; indeed, it could be deemed sufficiently elastic to include in the community of mutual aid those individuals with severe cognitive impairments.

VII. Paternalism and Legal Implications

So far I have argued that the Kantian account articulates a range of moral duties and provides plausible justification for third-party interventions in cases where an individual's agency has become disabled as a result of the first-order, relational circumstances. We have already seen how these claims provide an illuminating normative lens through which to interpret certain judicial interventions. But what about more controversial cases, where an individual's choice to remain within an autonomy-disabling relationship is overturned? The promotion of autonomy competencies might be deemed a worthwhile good amongst medico-juridical practitioners, but what if the individual concerned does not herself hold this value? Or let me phrase this objection in light of the discussion in the previous chapter. There I suggested that within the medico-legal space of reasons is a commitment to promoting an individual's autonomy competencies. The cases I focused on concerned instances of decisional incapacity resulting from relational contexts that disabled an individual's critical responsiveness to discrepancies between her own values and those within her primary-supportive relationships. Yet can we say that intervening in a disabling relationship is justifiable in cases where there are no clear discrepancies? The Kantian approach might track the deeper normative contours behind the SA and MM decisions, but whether it can justify overturning WMA's strong opposition to third-party intervention is more controversial. From a social model perspective, such third-party assessments of another's private relationships represent unwarranted paternalistic action founded on disabilism and discrimination: non-disabled individuals are free to remain in disabling relationships without third-party interventions – the same non-interference ought to apply to those with impairments. Passing judgment on relationships that fail to inculcate and promote autonomy competencies amounts to an unjustifiable imposition of foreign values onto individuals.

Two responses can be made. First, it is not clear to me that the ethical considerations in the case of WMA differ in substantial respects to those of SA and MM. In both sets of cases, these protective, dysfunctional relationships reinforced a crippling sense of powerlessness, making these individuals prone to other risks and vulnerabilities in the long term.[85] A further refinement of what we mean by vulnerability would be helpful in this context. According to Catriona Mackenzie, vulnerability can be discussed at three different levels. *Inherent vulnerability* refers to empirical facts of human vulnerability: that it is intrinsic and ineradicable to the human condition, given our embodiment, interests, and needs. *Situational vulnerability* is caused or increased by circumstantial and contextual factors, for example, marital breakdown. Finally, *pathogenic vulnerability* is morally questionable and results from objectionable practices, such as abuse, prejudice, or discrimination. The notion of pathogenic vulnerability further 'helps to identify the way that some interventions designed to ameliorate inherent or situational vulnerability can have the paradoxical effect of increasing vulnerability'.[86]

Within the Kantian framework, engaging in positive action, not just constraining or eliminating certain practices, ameliorates the risk of pathogenic vulnerability. Positive actions must start with a contextual understanding of the individual's particular impairments to help shore up their personal resilience and provide opportunities to realise their unique deliberative, agential potential. We might understand the normative focus in the cases of SA and MM as the provision of propitious conditions that enabled them to exercise their *already existing* autonomy competencies. By contrast, the judgments in WMA-like cases could be interpreted as seeking to provide opportunities for these individuals to develop *non-existent or weak* autonomy skills in the first place, which means a duty to remove them from relationships. In other words, *the failure to intervene at all*, as with a wrong type of intervention, can heighten pathogenic vulnerability. It is this more fine-toothed distinction that is lost in blanket prohibitions against intervening in relationships involving individuals with impairments. Respecting individuals as end-setters will morally require more than non-interference – or interference of the *right kind.* The dichotomy drawn between non-intervention and intervention is, in many respects, false, as discussed in Chapter 2. As we will see further in the next chapter, hermeneutics posits that every dialogical encounter is an intervention of some kind. But the important point for my purposes here is that interventions carried out in the right manner can express

[85] Catriona Mackenzie, 'The Importance of Relational Autonomy and Capabilities for an Ethics of Vulnerability', in Mackenzie et al., *Vulnerability*, p. 47; Dodds, 'Dependence, Care, and Vulnerability', p. 198.

[86] Mackenzie, 'The Importance of Relational Autonomy', p. 40.

inherent respect for the autonomous capacities of persons – perhaps even in a deeper sense compared with truncated definitions of respect premised on compliance with individuals' subjective preferences.

Consider, for example, the judicial decision to intervene in *A Primary Care Trust v P & Ors*, mentioned earlier.[87] Proposals to remove P to independent living were grounded on ameliorating the pathogenic vulnerability caused by the disabling views underlying his mother's intrusive care. His mother's perspective of P was revealing: she became 'deeply convinced that P was effectively an invalid with ME [myalgic encephalomyelitis or chronic fatigue syndrome]' even though this was contradicted by medical assessment. She subsequently 'fought to protect him from the intrusions of, as she [saw] it, incompetent, indeed mendacious professionals', effectively 'catapulting them down a vicious spiral of mutual interdependence which . . . resulted in each fulfilling the fantasies of the other'.[88] Thus, 'P appeared to the world as an epileptic and chronic sufferer of ME who was unable to socialise, dependent on a wheel chair, often and sometimes most of the day cared for by his inseparable and committed mother'.[89] Given this context of relational enmeshment, P's perspective in terms of planning for the future was 'severely restrict[ed]' and he could not 'visualise any prospect of having a different view to his mother on any subject that matters'.[90] In Hedley J's words, 'I feel that a return to AH will on the balance lead to the return of the pre-July 2007 position, with P being required to become sick, weak, and wholly dependent human being, to be protected at all costs from an intrusive and misguided state, in the shape of medical and care professionals, and to his being treated as AH and she alone thinks best'.[91]

Through the normative lens of my argument thus far, P's decision to remain with his mother could be construed as fundamentally undercutting his practical agency in the future. Thus, a Kantian perspective would sanction the intervention in this case on grounds that it constrains those maxims which severely limit the possibility of P developing those autonomy competencies that are constitutive of practical agency, thus disrespecting his status as an end-setting, temporally extended agent. P's removal was important for developing autonomy competencies and personal resilience in the long run, given that he would likely outlive his mother. It would create space between him and his mother to promote the acquisition of new experiences, and thus 'enable him to develop a

[87] *A Primary Care Trust v P & Ors* [2009] EW Misc 10 (EWCOP).
[88] Ibid., para. 26. [89] Ibid. [90] Ibid., para. 37. [91] Ibid., para. 67.

social and emotional independence . . . [free] from the dominating effects of an obsessive and smothering relationship', thereby developing genuine views of his own.[92]

I suspect that, even in light of the facts of this case and the argument thus far, proponents of the social model might maintain that P's wish to remain under his mother's care ought to have been respected given the presumption of universal capacity. But this position ignores how respect for autonomy presupposes that one *values* autonomy; and to value autonomy we need to value its necessary personal and relational conditions. Relationships that disregard the aspirational aspect of human agency ultimately create pathogenic vulnerability for individuals and endanger their capacity for temporally extended autonomous agency. The temporally extended character of our agency is not a feature we opt into or out of, but remains a fact of what it means to lead a life in light of one's own self-conception. Even though P himself might not have shared the goal of developing his own autonomy competencies and temporally extended agency, failure to do so would not only exacerbate both his situational and pathogenic vulnerability in a more fundamental sense, but signal a rejection of the type of agency that is presupposed in the prioritisation of autonomy as a value in the first place. Framed in Kantian terms, P's mother clearly recognised that her son was a dependent rational being. This recognition, however, was to the extreme, to the extent that his dependency was exaggerated at the expense of the aspirational dimension of human agency and social integration. Cases like P help illustrate how medico-judicial capacity adjudications can and should recognise the serious long-term harms that result from such an imbalance.

That said, my second response to the preceding objection comes in the form of a qualification. So far I have suggested that the medico-legal space of reasons affirms the value of developing an individual's autonomy competencies. Kantian theoretical resources help justify certain third-party, relational duties necessary to promote an individual's autonomy competencies largely external to the first-order, personal-supportive context. This suggests that third-party interventions can be warranted in cases of their neglect. However, none of what I have said indicates that this is a 'one-size-fits-all' moral solution. Particularistic, contextualised judgements are both crucial and necessary. Interventions in relationships that intensify pathogenic vulnerability tread a fine line between strong and weak paternalism and should proceed with caution. Ultimately, the

[92] Ibid., para. 64.

distinction between justified, enabling support and unjustified, disabling interference rests on *the ethical perspective and dialogical stance of the intervener*, particularly in cases where an individual's expressed wishes are overruled. Further normative content beyond the Kantian approach is required in terms of how to (i) enhance and promote an individual's autonomy competencies *within* relationship and (ii) prevent *external* third-party interventions from becoming disrespectful, strongly paternalistic interactions. In short, we still need a more substantive account of the ethical dialogical constraints both *within* and *outside* a supportive relationship, particularly if we are to sidestep concerns about discriminatory paternalistic treatment of individuals with impairments.

The WMA case helps illustrate my point. On one hand, ethical arguments help track reasons for intervening in WMA's desire to remain under the care of his mother. WMA's neglect was clearly noted in the case notes: their house was in such a filthy state of disrepair that it posed a health risk; outside help in terms of care and monetary assistance was regularly refused, resulting in WMA's social isolation and underdevelopment. His mother's dominance impeded his decision-making capacities and disagreement was not permitted.[93] It was alleged that his mother was 'psychologically abusive to her son, speaking about him in a derogatory manner'[94] and, like the case of P, he would be clearly ill-equipped in terms of key skills, competencies, and tools for everyday living should his mother pass away. In HHJ Cardinal's words:

> It is not property or making a will that matters in this case, it is WMA being able to cope, having life skills, socialising, managing a budget, providing for himself, skills he cannot and will not learn in any way, shape or form whilst living with his mother and without substantial local authority support. Would it not be better for WMA to move to an environment where he could learn these skills? But [WMA's mother] would not accept that. She conceded that upon her demise WMA would have to live in the social services accommodation. Was it not better for him to learn new skills now? She did not agree and that seemed to me to be entirely unrealistic.[95]

Moreover, an expert witness testified that WMA was a young man 'with the potential to lead a more fulfilling life', and his mother's 'inhumane and degrading' low expectations meant 'his true potential [had] been unrecognised and stifled'.[96] Construed through the lens of my account so far, assessments clearly indicated that the relational

[93] *A Local Authority v WMA & Ors* [2013] EWHC 2580 (COP), para. 26.
[94] Ibid., para. 15. [95] Ibid., para. 58. [96] Ibid., para. 67.

context neglected the development of WMA's autonomy competencies, exacerbated his pathogenic vulnerability, and undermined his end-setting potential. Based on what I have said thus far, intervention could be justifiable.

On the other hand, the judge's dialogical stance in the discussions of mental capacity and WMA's wishes and feelings is unintentionally problematic. The finding of incapacity ironically drew upon evidence by the very individual who allegedly inhibited WMA's social and intellectual development: '[e]ven [WMA's mother] has doubts as to his capacity and considers him less capable than others of achieving in this life'.[97] In addition, the judge took counsel that WMA's verbal articulacy should not be thought to be an indication of his borderline capacity in terms of deciding residence, care plan, and contact with his mother. HHJ Cardinal's further statements about capacity continued to reinforce the disabling presumption that WMA's autonomy competencies and skills were inherently limited, regardless of potential changes in relational context:

> I bear in mind section 4(3) requires me to consider if WMA may at some time have a capacity to make decisions as to residence, *et cetera*, but I take the view the evidence points unequivocally to the fact that, whereas he may improve as to his socialisation and skills, *he is most unlikely to regain or gain an ability to take a decision as to accommodation.*[98]

Rather ironically, then, the judge used the language of mental incapacity to emphasise the limited potential of WMA, an outlook that was itself an object of concern and underlined reasons for his removal from his mother's care. Even if HHJ Cardinal's finding of incapacity was legally correct, his statements reflected a contradictory ethical-dialogical stance according to my account of capacity. Paternalistic interventions made on grounds of promoting those autonomy competencies neglected in a disabling relationship must themselves be constrained by, and consistent with, the normative principles they are espousing: it seems contradictory to say that third-party interventions can be ethically warranted if one's primary relationship has failed to cultivate and develop one's potential capacities for autonomous decisional agency, but then adopt the attitude that these same inabilities will still persist, regardless of a different relational environment. The best interests decision could be similarly criticised on the same grounds that the dismissive attitude towards WMA's wishes and feelings reinforced a narrative of helplessness and incapacity.

[97] Ibid., para. 108. [98] Ibid., para. 115, second emphasis added.

Even with the recognition that WMA's wishes and feelings should be given greater weight in light of his borderline capacity and 'high level of functioning for a person with a learning disability', HHJ Cardinal argued that they 'were not rational, sensible, responsible and practically capable of sensible implementation' and therefore could not be accommodated in the best interests decision.[99]

The reasons behind WMA's removal from his mother are intuitively understandable. But this decision crosses the fine line between respectful, autonomy-promoting assistance and unwarranted paternalism, precisely because the ethical-dialogical stance reflected in the judgment is inconsistent with the justifying reasons underlying the intervention. Rather than challenge views about his limited potential, the disabling language used in both the capacity assessment and best interests decision reiterates views that exacerbate WMA's pathogenic vulnerability and undermine the promotion of his autonomy competencies. Medico-juridical practitioners rightly acknowledge that valuing and respecting autonomy sometimes warrants intervening into private relationships, precisely because the development of autonomy skills is vulnerable to, and dependent on, the conditions of one's relationships. But as this case illustrates, interventions must reflect a specific ethical-dialogical orientation, abiding by autonomy- and capacity-developing principles of conduct that sanction third-party involvement in the first place. Medico-juridical assessments function as an external form of moral judgement on one's primary-supportive relationships, but its dialogical practices remain *within* the same normative framework they are applying and should be judged accordingly. The fine line between strong and weak paternalism hinges precisely on this internal normative consistency. In scenarios where the promotion of one's autonomy competencies looks suspiciously like the imposition of a foreign value onto an unwilling individual, the ethical-dialogical standpoint of the third-party medico-juridical practitioner is all the more important. Such professionals are commonly considered *outside* a person's immediate relational context, yet they nonetheless become moral players in the relationship as soon as their mediating presence is used to overturn an individual's expressed wishes. Kant's theory can provide a partial normative framework for these interventions, but, as we will see in the next chapter, further substantive content is required to flesh out these constraints properly.

[99] Ibid., paras. 21, 124.

Conclusion

I started out this chapter with a challenge to the social model of disability and care theory and their theoretical capacities to justify the positive duties owed to individuals with impairments. More robust theoretical resources are needed to account for the complex interaction between impairment, disabling relationships, and pathogenic vulnerability, as well as its mitigation through the relational duties of support. I have suggested that an account of Kant's Formula of Humanity that begins with premises not of equal dignity and respect for autonomy, but of unavoidable human interaction, interpersonal dependency, and rational contingency, can provide a normative framework for us to explain and justify the range of interpersonal duties of respect and support owed to individuals with impairments. Kant's derivation of both positive and negative duties stems from a dual recognition of the importance of striving towards an ideal of end-setting agency, but also the intrinsic relational constituents and dependencies that inhibit individuals from achieving this ideal in isolation. Medico-juridical cases illustrate how relational oppression and neglect often result when the aspirational dimension of agency is unbalanced with the recognition of one's limitations, and vice versa. Filling in this gap are positive duties of assistance to help develop and promote an individual's autonomy competencies. These duties not only guide the practices within primary-supportive relationships but, more crucially, indicate that third-party interventions are morally permissible and may even be obligatory.

But this is only a partial account thus far. These interventions must also be subject to certain normative constraints. For this we still need to clarify the ethical-dialogical constituents of those interactions that respect and promote individuals' potential autonomous competencies. As the next chapter explains, the concept of 'hermeneutic competence' can capture the ethical practices intrinsic to autonomy-enhancing interpersonal and dialogical interaction, both internal and external to primary-supportive relationships.

6

Hermeneutic Competence and the Dialogical Conditions of Capacity

So far, I have outlined crucial facets of a relational conception of capacity, such as relational rights, autonomy, and rationality. Relationships supportive of decisional capacity promote various skills of absorbed coping and autonomy. Moreover, medico-juridical assessments are situated within the social space of reasons and need to be conscious of an articulatory-disclosive mode of reasoning operating implicitly when they apply the functional test of capacity. Invariably capacity adjudications articulate and affirm the worth of certain values, highlighting in particular the importance of individuals' expressed self-understanding. Such self-understanding functions as the point of entry for third-party, normative appraisals of an individual's primary-supportive relationships. I further established that the question of third-party obligations requires a more relational concept of rights that, when combined with the Kantian understanding of the dynamic between aspiration and ability, may sanction positive duties of assistance and support in relational contexts that undermine an individual's decisional capacity.

The purpose of this chapter is to elaborate on the intersubjective skills and abilities that are required to enable the decisional capacity of individuals with impairment. Chapter 3 brought the constituents of autonomy competencies to the forefront, such as perceptual skills of absorbed coping and reflective capacities of self-understanding from the first-personal perspective. I want to shift the normative focus away from the individual in question. If capacity is truly a *relational* concept, then we need to consider the skilful dialogical practices *internal* and *external* to one's primary-supportive relationships that promote the autonomy competencies of individuals with impairments. There are dual impulses in the treatment of those with impairments: to assimilate them to the extent that they become subsumed or merged with another or to treat them as totally other, to the point that they may be viewed as completely incomprehensible, outside the realm of our understanding, and dehumanised as a result. This can occur in both contexts.

What I call *hermeneutic competence* is crucial if we are to mitigate these dual dangers. Hermeneutic competence is an interpretive, dialogical orientation that is grounded in recognitional mechanisms based around moral consideration owed to the unique separateness of individuals, regardless of their impairment and disparate levels of requisite care. To be clear, my term hermeneutic competence overlaps with, but also differs from, a *hermeneutic approach to competence*: the latter articulates how assessments of capacity should be focused on enabling individuals to express themselves, their sense of self, values and meaning, through dialogical interaction with others.[1] The account offered here is consistent with this approach to capacity. However, what I call hermeneutic competence refers more to the substantive background capacities and abilities of persons *surrounding* individuals with impairments, abilities that are in fact *presupposed* in a hermeneutic approach to competence. In this respect, hermeneutic competence implies a significant ethical reorientation, where the focus of competency shifts *away from individuals with impairment*, and *towards the skills and competencies of those who are responsible for responding to their needs and capacities.*

Section I challenges prevalent dichotomisations of care and respect and illustrates how relationships premised on asymmetrical responsibility and unequal capacities demand hermeneutic competence. An important constituent of hermeneutic competence will be phenomenological awareness, as I discuss in Section II. Relationships must be attuned to the first-personal, phenomenological experience of absorbed coping discussed in Chapter 3 in order to establish a shared perceptual understanding that is necessary to support deeper connections between individuals. This shared perceptual understanding helps validate a person's experience of skilled coping and everyday autonomy. The recognitional dialogical practices that promote shared understanding are examined more carefully in Section III. Power dynamics in relationships characterised by asymmetrical responsibility pose a genuine risk, where another's impairment is exploited to damage his or her practical identity

[1] Lazare Benaroyo and Guy Widdershoven, 'Competence in Mental Health Care: A Hermeneutic Perspective', *Health Care Analysis* 12:4 (2004): 295–306. Benaroyo and Widdershoven likewise argue that a hermeneutic approach 'requires that the evaluator is competent (in the sense of practical rationality) herself' (p. 300). I set out more specifically what competencies are required by those around the person whose capacity is in question, and these rely on skills beyond practical reasoning, though such skills are obviously implicit in my discussion of dialogical ability and interpersonal recognition.

and entrench oppressive, disabling practices through objectifying and assimilating dialogical practices. Recognitional dialogical practices, by contrast, are grounded in reflexivity, humility, and openness towards the challenging views of others.

Section IV explores how interpersonal recognition cultivates nurturing relations-to-self. Individuals come to mirror the values and ways in which they are treated and recognised, so that these are internalised and perpetuated as a way of relating to oneself. Comprising self-respect, self-trust, and self-esteem, such relations-to-self are preconditions for the development and exercise of autonomy skills. In this way, engagement and dialogue with others can often have a simultaneously *constitutive* and *reflective* function for those with impairments. Section V applies the constituents of hermeneutic competence to the unique interactions and specific normative function of capacity assessments. The previous chapter emphasised that the fine line between intrusive, illegitimate intervention and positive duties of support depends on the ethical-dialogical orientation of the intervener. I examine this orientation in more detail in Section VI, arguing that recognitional dialogical practices are important for justificatory consistency and to prevent overdetermination in capacity assessments. Moreover, these practices will help assessors identity the need for, and instantiate the process of, narrative repair which bolsters individuals' resistance to disabling ways of understanding themselves and develops enabling counter-narratives as an alternative.

I. Asymmetrical Responsibility and Hermeneutic Competence

The discussion thus far has scrutinised the theoretical concepts that underlie the medico-juridical interpretation of decisional capacity, steering clear of deeper issues around what this implies about the relative moral standing of individuals with or without capacity – namely what kind of treatment is *morally owed* to individuals, whether it be respect for one's choices or paternalistic best interests decision-making on one' s behalf. From the law's perspective, respect for choice hinges on capacity, the evident display of certain functional skills. From the perspective of prominent moral theories, such as contractualism, respect for one's choices hinges on something deeper, such as the ability to be held accountable to others for one's actions. We might initially think of capacity as separate from the question of moral responsibility in the latter sense; indeed, this seems to be the dominant medico-juridical approach.[2] With closer examination, however, the legal distinction between capacity and best interests is premised on *different ethical orientations* that are required in relationships involving individuals with unequal capacities. And it is here that contractualism's account of moral responsibility is particularly illuminating.

According to contractualism, different orientations of *respect* and *care* are founded on a threshold concept of moral responsibility: if individuals within a relationship pass a threshold of some kind of competence, they can be held morally responsible for their reasons and actions. They are each bound by respect for the agency of the other, expressed through deference to a person's subjective choices.[3] In short, only individuals who possess the psychological competence to engage in equal moral responsibility are owed respect. More sophisticated versions of contractualism situate respect within the second-person standpoint – in the sense that the source of moral obligation issues from another's demand – but still presuppose the possession of certain psychological competencies for which those with mental impairments may fall short.[4] The upshot is that such individuals may be subject to paternalistic care rather than owed respect for their autonomy. Whether duties of respect or care are owed to an individual ultimately depends on the competencies of the individual with impairments.

This way of framing respect and care is deeply influential, both in theory and practice. My focus here is not to explore contractualism and its implications, however.[5] The important point I wish to make is that even as mental capacity law seems to eschew the language of moral responsibility, its division between capacity and best interests assumes a strikingly similar dichotomisation between respect and care. Duties of respect are

[2] This clearly is not the case in criminal law, however, where mental competence or incompetence can alter the degree to which one is responsible for a crime. Moreover, in *LB Haringey v FG* [2011] EWHC 3932 (COP), Hedley J appeals to the concept of responsibility as a factor in determining the extent to which incapacitated individuals can participate and determine the best interests decision of 'whether the person has responsibility for making and living with the consequences of any decision which they choose to make' (para 15).

[3] For more on this debate, see Darwall, *Second-Person Standpoint* and *Welfare and Rational Care*, as well as T. M. Scanlon, *What We Owe to Each Other* (Cambridge, MA: Harvard University Press, 2000) and his book *Moral Dimensions* (Cambridge, MA: Harvard University Press, 2010). I discuss both in Kong, 'Space between Second-Personal Respect and Rational Care'.

[4] Kong, 'Space between Second-Personal Respect and Rational Care'.

[5] I discuss this more fully in Ibid.

owed to you if you have capacity; duties of paternalistic care apply if you are found incapacitous. But as we have seen thus far, this allows for no middle ground for those who lack capacity but should nonetheless have their decisions deferred to, and for those who might have capacity yet are ensconced within fundamentally uncaring, abusive environments. Likewise, contractualism's emphasis on equality and reciprocity can at times overdetermine the mode of engagement owed to individuals with impairments. Equal respect for one's autonomy demands equal moral responsibility – that one's justificatory reasons for choices and actions can withstand the scrutiny and autonomous consent of others within the contractualist circle. But if individuals with psychological impairments of some kind cannot assume this form of mutuality and equal responsibility, it seems to follow that their claim to equal respect is likewise in jeopardy. This unpalatable result could morally sanction an overly paternalistic orientation towards those with impairments or, at the very least, distort important nuances in the type of deliberative respect and recognition that is owed to them.[6] Contractualism's problematic dichotomisation of respect and care thus brings to light the neglected middle ground between capacity and best interests, particularly when we start to consider decisional capacity as situated within enabling or disabling relational contexts.

Care and respect cannot be easily separated when we speak of relational practices that enable the agency of individuals with impairments, even as asymmetrical responsibility may characterise these relationships. One response to contractualism might appeal to theories of care that reject respect for autonomy as the preeminent moral value and fully recognise the interconnected, caring strands that help us flourish. But as we saw in the previous chapter, care theory idealises empirical facts about human fallibility and dependency, with the intent on reinforcing the universal experience of care; yet this disregards the phenomenology of unequal moral responsibility that often characterises relationships involving individuals with impairments. This makes it too indeterminate to articulate the complex interplay between care and respect in the relational practices that enable the decisional capacity of those requiring greater support.

We can better understand the dynamic of respect and care through the concept of mutual recognition, an intersubjective process through which we begin to see ourselves in a certain light based on the status accorded to us by others. Indeed, as Levinas implies, recognition can occur even in conditions of unilateral moral responsibility. Levinas describes

[6] Ibid.

how exposure to another's vulnerability binds me to a categorical ethical demand of non-indifference to her. Specifically, experiencing another's vulnerability and dependence elicits a sense of unilateral rather than reciprocal moral responsibility. This primordial ethical summons comes prior to any verbal expression of a demand and captures the moment when one recognises a 'relation to the absolutely weak – to what is absolutely exposed, what is bare and destitute'.[7] Non-indifference is steeped in recognitional bonds: it denotes the *objective recognition* of radical dependence and asymmetrical responsibility, as well as the *intersubjective, agent-relative recognition* of another's unique separateness and independence from me, their vulnerability to my actions, my neglect.[8] Even within a relational framework of asymmetrical responsibility, respect for the unique perspective of the other is ethically required.

At this point, the difference between a contractualist position and a recognitional framework based on Levinas might appear nominal: the former would likely agree that we need to 'recognise' individuals in some sense, even if this means that we do not accord the same moral status to their choices. They would also agree that individuals with impairments are owed moral obligations and that others are responsible for fulfilling them. But the key difference is that the contractualist framework *already determines* the *content* of those obligations, the *mode of* engagement, and *form of* recognition that is normatively required, *based on whether individuals fall short of the threshold of equal moral responsibility*. If they do, they are owed duties informed by objectivist, welfarist judgements – essentially best interests decisions. One is no longer recognised as a *subject*, but viewed as an object of concern – or worse, an object to be governed, managed, and controlled, if others deem this as appropriate to one's welfare or best interests. There appears to be no middle ground between the respect that is owed to individuals and their choices (so long as they meet a certain competency threshold) and the third-party paternalism that is conferred onto individuals should they fall short. Certain accounts potentially try to address this middle ground – to capture the kind of respectful recognition that is owed to individuals *even should they fall below the threshold*. Ultimately, however, these theories have to undergo unconvincing philosophical manoeuvring to try and fully accommodate the intuition that *even when* individuals might lack

[7] Emmanuel Levinas, 'Philosophy, Justice, and Love', in *Entre Nous*, trans. Michael Smith and Barbara Harshav (London: Continuum, 2006), p. 89.
[8] Ibid.

competency, they are owed the type of respect that characterises those judged to have equal competence.[9]

By contrast, the import of Levinasian observations about recognitional but asymmetrical relationships is that it resists the temptation to predetermine the content of one's moral obligations, mode of engagement, and form of recognition that is owed to individuals with unequal competence. The ethical focus shifts away from the vulnerable or fragile Other – where it is *her* competency that determines the mode of engagement that is owed to her – and towards the competency of the person *on the receiving end of the ethical summons.*[10] The recognition of radical human frailty and dependence can prompt unilateral responsibilities that require the recognition of another's alterity, without determining the status and capacities of individuals in the first instance.[11] For example, consider the relationship between a parent and her severely impaired child who cannot communicate through conventional means.[12] These vulnerabilities express the child's dependency in an acute manner, activating the parent's categorical responsibility to the child, regardless of whether the latter can reciprocate in kind.[13] Crucially, however, such asymmetry need not invalidate the recognitional bond from parent to child, nor determine the actual mode of engagement between the two individuals. The severely impaired child clearly lacks the same psychological capacities of her parent – or indeed, of children of her own age. She may not be able to reciprocate in terms of conferring status, or recognising the parent as an object of concern and contributor to certain interests and projects. Conversely, the identity of the parent may not be vulnerable to the child's ascriptions of her to the same degree. Her relationships take on a non-reciprocal

[9] I discuss this at length in relation to Darwall and Scanlon in ibid.

[10] Ibid., pp. 456–66. There I discuss how exposure to vulnerability leads to a sense of categorical responsibility.

[11] This is an important difference from care theory, which likewise posits the priority of responsibilities, but then determines the specific mode of engagement.

[12] Indeed, a parent has unilateral responsibility to his or her child without impairments, simply by virtue of being a parent. As the child matures, however, it may be the case that the relationship develops some kind of reciprocity: for example, a parent financially supports the child until she might have developed skills to make a livelihood in the world – not all parents feel obliged to pay a deposit for housing for their child (they might think the child is morally responsible for their financial/livelihood choices), but all parents are obliged to house their children before they are legally adult. The eventual transition to a more reciprocal relationship may be slower or may not apply in the case of a relationship between parent and severely impaired child. My thanks to Tony Hope for pushing me to clarify this example.

[13] Kong, 'Space between Second-Personal Respect and Rational Care', p. 458.

structure in terms of moral responsibility, and indeed, what we might call, recognitional responsibility (or the duty to recognise an other). Yet the facts of unequal competency or non-reciprocity by themselves do not lessen the necessity of establishing recognitional bonds that fundamentally treat the child as a *subject* (as a separate being with unique needs or perspective), nor do they determine as of yet whether paternalistic care or respect for her choices is owed to her; they do not in themselves dictate the type of engagement that is appropriate in the circumstances. It is possible that when the parent overrides her child's expressed wishes, she still does it in a way that fundamentally confers deliberative respect for the child's own views. Conversely, the parent might very well accord respect to the child's choices that go against what the parent believes to be in her best interest – not because the child has equal competence or moral responsibility, but because the parent recognises the child as a separate person from her, where respect for this choice shows a fundamental trust in her child and her abilities. Indeed, even as asymmetry characterises this relationship, it is not clear that it can legitimately be characterised as 'non-reciprocal' (as we will see later).

To clarify, Levinas' theoretical framework on its own cannot provide concrete guidance in terms of what is needed in certain relational contexts. But it does help clear space for a rich middle ground where care and respect, autonomy and concern for another's welfare, can potentially intersect in recognitional bonds. Appropriate expressions of this middle ground require that the receiver of the ethical summons has *hermeneutic competence*: what I define as the moral attunement to attend to another individual through dialogical skills and virtues, even under conditions of unequal, non-reciprocal moral responsibility. Such competence helps establish recognitional bonds characteristic of enabling relationships and expresses an outward, interpretive orientation that facilitates the attenuated understanding, care, and deliberative respect towards another. Ultimately, hermeneutical competence helps dismantle the strict dichotomy between respectful and caring modes of engagement – mainly because the principal question is not oriented around 'what conditions do we defer to another's subjective choices?' or 'how do we express care in this situation?' These questions do not capture the full sense of what it means to respect the separateness of persons. Hermeneutic competence focuses on the recognitional, dialogical conditions that not only enhance or detract from decisional capacity, but influence how meaning is conferred onto an agent's actions or words, making it thereby crucial for those both inside and outside individuals' primary relationships. In short, it directly affects both the *internal practices* as well as *external interpretation* of capacity.

Three components comprise hermeneutic competence: (i) *phenomenological awareness* denotes attentiveness to the broader everyday conditions of absorbed coping. Responsiveness to facts of embodied vulnerability will improve our interpretive understanding of individuals' ways of coping with perceptual constituents of experience. (ii) *Dialogical openness* involves a stance of humility, where recognitional modes of engagement are open to the testing of one's views and assumptions, thus overturning presumptions that one knows what is best for another. Finally, (iii) *nurturing relations-to-self* refers to the modes of interpersonal recognition that lead to the cumulative internalisation of autonomy competencies, such as self-respect, self-trust, and self-esteem.

II. Phenomenological Awareness

Phenomenological awareness attends to how impairment impacts on perceptual constituents of embodied coping (specifically, one's experience of space and time). Spatial disorientation can emerge in cases of bodily restrictions or when unfamiliar, impersonal spaces come to dominate our experience, thereby heightening our sense of isolation. As William May describes:

> Successively and progressively, impairment, old age, immobility, and death restrict space. The world at large shrinks to a single room and ultimately to a casket. Ordinarily, people live in a number of different environments – home, workplace, streets, parks, gardens, and sidewalks. The bedroom is only part of a total world, often a sanctuary from it. But, for the immobile or impaired, the world contracts to a single room. Designers of total institutions take on an awesome responsibility. They create for residents not just a fragment but the whole of their perceivable world.[14]

Impairments can constrict the subjective meaning and perception of space. George Agich explains, for example, how wandering or confused behaviours amongst elderly long-term care residents typically elicit care responses of 'management' or 'control' through the use of physical

[14] William May, 'The Virtues and Vices of the Elderly', in Thomas R. Cole and Sally A. Gadow, eds., *What Does It Mean to Grow Old? Reflections from the Humanities* (Durham: Duke University Press, 1986), p. 46, quoted in George J. Agich, *Autonomy and Long-Term Care* (Oxford: Oxford University Press, 1993), p. 126.

restraints and alarms.[15] Even as wandering behaviour might be concerning, phenomenological awareness makes us attuned to how it is nonetheless an understandable coping strategy responsive to subjective experiences of spatial distortion, aggravated further by diminishing sensory-motor skills. A phenomenologically attuned response would recognise the 'latent meaning in the seemingly purposeless movement', to provide a 'purposeful routine' of such behaviour, to manipulate physical barriers in order to facilitate individuals' mobility and coping engagement with their environment.[16] Thus, attention shifts towards adapting spaces to help establish a shared experiential world with those who have impairments.

Impairment can likewise disrupt an individual's ability to adapt to and function within unfamiliar ways time is structured. Naoki (whose book was first discussed in Chapter 3) describes his experience of time, for example:

> For us, one second is infinitely long – yet twenty-four hours can hurtle by in a flash. Time can only be fixed in our memories in the form of visual scenes. For this reason there's not a lot of difference between one second and twenty-four hours. Exactly what the next moment has in store for us never stops being a big, big worry.[17]

When we are insensitive to another's perception of time, we misunderstand why their daily rhythms, rituals, or schedule might depart from one's expectations. This applies whether or not an individual has impairments. To use a mundane example, I might not notice how quickly time has flown when absorbed in a task until my husband reminds me, 'It's 7 PM! We're late for our dinner date!' I might then be anxious about the fact we're late or annoyed that I lost track of time, but none of these responses will lead me to think I'm fundamentally out of sync with the abstract measure of time that I share with my husband and broader community. Yet sometimes impairment can alter one's sense of time in profound ways. Naoki's description of his anxiety seems to articulate something deeper, where one's core perceptual experience seems disconnected from that of others – i.e. the very manner in which time is divided up (albeit a construct but a common, socially accepted one) departs

[15] Agich, *Autonomy and Long-Term Care*, p. 130. This is not to dispute the fact that wandering could potentially be dangerous, both for the individual concerned and other residents. But the point I want to make is that these attempts to 'control' such behaviour may stem from a misreading of the purpose behind such actions, and further aggravate the distress and unease motivating it.

[16] Ibid., pp. 125, 129–31. [17] Naoki, *The Reason I Jump*, p. 96.

significantly from one's own temporal experience – such that it reinforces a sense of discomfort, of not fully belonging in the world. There are very ordinary examples where this occurs. Highly bureaucratic, institutional care regimes might impose routines and schedules of care-recipients at the expense of their particular needs or expectations to accommodate the workday of staff.[18] For example, a physically impaired individual requiring help to bathe might have a preference for this to occur at night as part of her own process of winding down in preparation for bed. But the limited scheduling of the care facility might stipulate bathing has to be done in the morning. At root, a lack of common time interrupts an important mechanism through which we understand another. When it is assumed that *our* experience of time is exactly the same as others who require assistance, it appears as though some kind of foreign temporal structure is being imposed onto them. We then disrupt their unique form of embodied coping, their particular search for equilibrium and ease within the world. Naoki, for example, depicts his own experience of time 'as difficult to grasp as picturing a country we've never been to.'[19] He encounters problems with seeing a visual schedule in the following way:

> People with autism may look happier with pictures and diagrams of where we're supposed to be and when, but in fact we end up being restricted by them. They make us feel like robots, with each and every action pre-programmed. . . . The message I want to get across here is: please don't use visual things like pictures on our schedules because then the activities on the schedules, and their times and timings, get imprinted too vividly onto our memories. And when that happens, we end up stressing ourselves over whether what we're doing now is or isn't matching up with what was on the schedule. In my case, I end up checking the time so often that I'm no longer able to enjoy what I'm doing.[20]

Such temporal incompatibility potentially reinforces the sense of constricted agency amongst individuals with impairments. It can exacerbate another's anxieties and sense of isolation, limiting their self-expression and agency as a result.

Disruption of common time doesn't just have mundane everyday consequences. Consider how time helps orientate one's subjective narrative history; it reminds us of our historical situatedness, in terms of our biography and traditions. For example, the children of a dementia patient might very well know that their mother became a speech therapist after

[18] Agich, *Autonomy and Long-Term Care*, p. 138.
[19] Naoki, *The Reason I Jump*, pp. 95–6. [20] Ibid., pp. 147–8.

her children grew up, mainly to help ease the financial burden of their family at the behest of their father's demands. But if her biographical narrative changes to her becoming a speech therapist before her children were born as a way to make a stand about the equality and potential of women in certain professions, in some respects it doesn't matter whether she is objectively correct about the temporal structure or the detail of her original motivations. Phenomenological awareness amongst those around her would recognise the crucial orientating function of these subjective narratives, of sharing her sense of time, whether or not they cohere with their own memories or are in fact true. 'Reconstructing biography and telling stories of one's life are integrative efforts leading to an overall intelligibility,' Agich explains. 'Stories [about oneself] provide an . . . orientating function that is widely employed in everyday life'.[21] Disregarding the phenomenological importance of shared time can inflict deeper damage to one's identity, particularly when considered in light of the self-constituting function of past narratives for ourselves and others. When these narratives are dismissed, it disrupts the process of self-orientation – whereby an individual's identity is aligned with temporal dimensions that make subjective sense.[22] Like space, sensitivity to another's sense of time therefore has a similar potential to establish a shared world with individuals with impairments, contributing to their sense of social belonging and integration.[23]

To summarise, phenomenological awareness makes us more attuned to how impairment may lead to divergent experiences of perceptual conditions, such as space and time, which alters individuals' expression and manner of absorbed coping accordingly. Recall that absorbed coping helps one discern how to move or act, whether this be triggered through some gut, bodily, or relatively non-intellectual feeling. Phenomenological awareness is a necessary (though not sufficient) condition to facilitate understanding of the coping behaviours one might find bewildering – for example, when an elderly person wanders in residential care, when the biography of a person with dementia changes, when an individual with autism takes a long time to get from A to B, yet completely panics when his schedule changes at the last minute. Through the lens of absorbed coping, these types of everyday acts form a response to familiar or unfamiliar organisational features of perceptual experience and could be interpreted as consistent with everyday expressions of autonomy – where an individual draws upon her reparatory of autonomy competencies to

[21] Agich, *Autonomy and Long-Term Care*, p. 139. [22] Ibid. [23] Ibid., p. 134.

perform particular acts that are consistent with her personal identity 'in the midst of things'.[24] Such phenomenological awareness helps initiate relational practices that seek to *understand* and *recognise* an individual's unique mode of engaging and coping with the world. Such awareness will therefore form the foundation for any attempts to cultivate another's autonomy competencies, particularly in the context of impairment.

The presence or absence of phenomenological awareness can directly affect another's social existence and practical agency in enabling or disabling ways. Consider two cases that revolved around the residency of elderly individuals. In *Re GC and Anor*[25] it was decided that, despite worries about the squalid conditions he was found living in, an elderly gentleman should be permitted to return home to his nephew on a trial basis, given his long-standing wishes. Hedley J stated, 'for the elderly there is often an importance in place which is not generally recognised by others; not only physical place but also the relational structure that is associated with place. Those seem to be the matters that are important to GC and underpin the expressions of wishes and feelings that, as I say, he has consistently advanced in this case'.[26] *KK v STCC* similarly concerned an elderly woman with vascular dementia and hemiplegia who wished to return home following her transfer to a care facility. Unusually for the Court of Protection, KK provided both written and oral evidence, citing perceptual dimensions that were important for her everyday experience of autonomy. She described her home as reflecting her own narrative history: '[e]verything I've got is in that bungalow. My whole life. Everything there is familiar to me. I've got my hobbies. I've got all sorts of things.'[27] KK's spatial and temporal experience seemed literally to shrink when comparing her home and her room at the care home.[28] From the vantage point of her bungalow window, KK said she could see 'everything in the village . . . – the church and the tower, the whole village'. By contrast, she 'tend[ed] to sit in her room' in the care facility.[29] In the same

[24] Agich, *Autonomy and Long-Term Care*, p. 100.

[25] *Re GC and Anor* [2008] EWHC 3402 (Fam).

[26] Ibid., para. 21.

[27] *KK v STCC* [2012] EWHC 2136, para. 42.

[28] As Agich states, this is more usual than not: 'It is not simply that the environment or setting changes as an elder moves from an upstairs bedroom to a first-floor living room, from an apartment into a child's home or an institutional long-term care setting, or from custodial to skilled care within a nursing home, but the significance and meaning of these changes of scene are just as pregnant with possibilities and fraught with peril as scene changes in a murder mystery.' *Autonomy and Long-Term Care*, p. 128.

[29] *KK v STCC* [2012] EWHC 2136, para. 42.

vein, KK responded to concerns raised about inadequate nutrition and hydration should she return home: '[i]f I was to return to my bungalow I would look forward to planning my meals and writing a shopping list with carers.'[30] She referred to the food at STCC (the care home) as 'baby food' and stated, 'I get frustrated that STCC's staff mash my food up and give me a spoon to eat it with. I do not need my food mashed up or a spoon to eat with.'[31] The importance of shared sense of time is implicit in KK's considerations about her future care needs: 'Usually I go to bed at 1900 hours and wake at 6 o'clock. Prior to my transfer to STCC I was put to bed by carers at approximately 1900 hrs and was visited again at approximately 6 o'clock at which time they would wash and dress me and put me in my recliner chair. This worked well.'[32]

The local authority claimed that KK's inability to recall her anxiety at being alone in her home – evident in her excessive overuse of 'lifeline' emergency call service – indicated her lack of capacity to use and weigh information.[33] But in his finding of capacity, Baker J contended that 'the local authority ha[d] not demonstrated that it has fully considered ways in which this issue could be addressed, for example, by written notes or reminders, or even by employing night sitters in the initial stage of a return home.'[34] Notes and reminders were thought to likewise address concerns about KK's nutrition and hydration, where 'more could be done to address this issue . . . by paying greater attention to KK's likes and dislikes'.[35] Baker J understood KK's testimony in the following manner:

> In weighing up the options, she is taking account of her needs and her vulnerabilities. On the other side of the scales, however, there is the immeasurable benefit of being in her own home. There is, truly, no place like home, and the emotional strength and succour which an elderly person derives from being at home, surrounded by familiar reminders of past life, must not be underestimated. When KK speaks disparagingly of the food in the nursing home, she is expressing a reasonable preference for the personalised care that she receives at home. When she talks of being disturbed by the noise from a distressed resident in an adjoining room, she is reasonably contrasting it with the peace and quiet of her own home.[36]

It is notable to compare the interpretive standpoint of the judge and the patronising inferences made at times on KK's behalf in her immediate care environment. Both recognised that KK's excessive use of the lifeline

[30] Para. 43. [31] Paras. 53, 43. [32] Para. 47.
[33] Para. 30. [34] Para. 71. [35] Para. 72. [36] Para. 70.

were indicative of anxiety at home.[37] But Baker J's judgment targeted ways in which her care could be altered to attend to KK's phenomenological experience of absorbed coping and everyday autonomy.

Cases like KK illustrate the importance of phenomenological awareness in relational conditions of asymmetrical responsibility: neglecting perceptual experiences of space and time can erect barriers to the understanding and interpretation of embodied coping behaviour, accentuate confusion and anxiety, or restrict the scope of and potential for another's practical agency. Attunement to these features, however, establishes important bonds of commonality which, as we see below, underlie the enabling dialogical practices of autonomy-enhancing care relationships.

III. Dialogical Understanding

At the heart of hermeneutical competence is a normative concept of dialogical understanding, which presupposes a recognitional, interpretive stance that incorporates self-reflection, humility, and deliberative respect. Disputes in capacity adjudication often attest to the challenge of achieving such dialogical understanding, whilst instances of undue influence and abuse represent its complete absence. Gross negligence of care results all too often when we misunderstand, minimise, or dismiss its importance, as the shocking case studies in Mencap's *Death by Indifference* reveal. There is a lot riding on this concept of dialogical understanding as a result. I want to discuss this feature of hermeneutic competence in some depth, mainly because it is often this that requires most skill on the part of the relationships surrounding individuals with impairments.

Dialogical understanding refers to the *interpretation and conferral of meaning* to another's words and actions. This is meant in a deeper sense than merely interpreting another's expressions or statements charitably in order to make them comprehensible to ourselves, based on our own frameworks of meaning.[38] Dialogical understanding demands critical scrutiny of our own prejudices and assumptions. We all are steeped within 'conceptual frameworks' or horizons of meaning – essentially

[37] But interestingly, the use of the emergency lifeline could also be interpreted as a form of absorbed coping for interaction with outsiders. This is a particularly salient point when we consider the number of elderly who suffer from deep loneliness and isolation. Thanks to Alice Obrecht for making this point.

[38] See Donald Davidson, 'On the Very Idea of a Conceptual Scheme', in *Inquiries into Truth and Interpretation* (Oxford: Clarendon, 1984) and Taylor's critique, 'Understanding the Other: A Gadamerian View on Conceptual Schemes', in *Dilemmas and Connections*.

perspectival orientations from which we can 'see'. These horizons are not value-neutral, however. Pernicious, oppressive influences that impede the development of personal identity and autonomy are sometimes concealed in linguistic practices.[39] A Janus-faced orientation is needed to root these out: an open dialogical stance towards others will mean we are likewise open to scrutinising our own existing prejudicial horizon.

Critically engaged, dialogical understanding is both a virtue and achievement, but not in the sense that we typically assume. Oftentimes critical understanding idealises a neutral, more objective point of view that is achieved by disengaging from our situated perspective. But this description distorts the process through which our understanding becomes enriched. Gadamer's hermeneutics show that the concept of prejudice is key to the process of dialogical enrichment. Prejudice denotes our surrounding intersubjective, cultural, and historical context that constitutes 'the horizon of a particular present'.[40] It forms our interpretive standpoint, similar to how a visual horizon dictates our range of sight from a particular but unfixed vantage point. Our prejudicial standpoint appropriates a dominant meaning and determines our interpretations, consciously or unconsciously. Indeed, language or analogous communicative conventions embed the unconscious influence of prejudice. Language functions as a repository of value and meaning, representative of our belonging to particular traditions. When we assume that understanding is the achievement of a value-neutral perspective, we dismiss our situatedness within a particular tradition. We therefore deny ourselves the material through which our perspective can grow, develop, and undergo criticism. A 'tyranny of hidden prejudices' results, entrenching our particular standpoint and foreclosing our critical engagement with it.[41]

To illustrate with a very mundane example, individuals of certain ethnicity might encounter the following situation: I might meet a person who looks different (i.e. not-Caucasian) from me. I ask her, 'Where are you from?' And she replies, 'I'm from Canada.' But then I reply, 'No, where are you *really* from?' And she responds, 'I'm *really* born and raised in Canada.' In the first instance, it is clear that I possess certain prejudices – I

[39] Hilde Lindemann Nelson, *Damaged Identities; Narrative Repair* (Ithaca: Cornell University Press, 2001).

[40] Hans-Georg Gadamer, *Truth and Method*, 2nd ed., Trans. Joel Weinsheimer and Donald G. Marshall (London: Continuum, 2004), pp. 304–5.

[41] Gadamer, *Truth and Method*, p. 272.

immediately presume that because the person looks different from me, she must speak English differently from me, she must come from somewhere else, and so on. This *is* my inextricable platform from which I engage with this person in the first instance, and perhaps there are reasons for it: I might have grown up in a predominantly white neighbourhood, had no exposure to individuals of different ethnicity, and someone who looks different might be a bit intriguing to me. It doesn't even become inherently problematic (though it might be quite annoying for the other individual) when I ask the second question, as it might indicate that I am genuinely trying to grapple with a new perspective I simply have not yet encountered. However, the 'tyranny of prejudice' might be revealed in the third question if I were ask, 'Okay then, where do your *parents* really come from?' rather than use this new perspective to question my own previously held prejudgements about the absence of non-Caucasian individuals living in Canada.

At this point, we might wonder if the cultural relativists are right: ensconced as we are within our own prejudicial frameworks, shared dialogical understanding seems even more unachievable. The more appropriate goal is the deconstruction and fracturing of frameworks, not fostering greater understanding of them. But this misunderstands the normative potential in prejudice. As Dreyfus and Taylor rightly argue, 'in dealing with real, partial barriers to understanding, we need to be able to identify what is blocking us. And for this we need some way of picking out the systematic differences . . . without either reifying these differences or branding them as ineradicable.'[42] This is important for both the skilful dialogical practices internal and external to one's primary-supportive relationships: in certain circumstances, individuals with impairment might initially appear to be expressing themselves or communicating with others in bewildering, incomprehensible ways. Establishing communicative bonds requires one to identify one's prejudice in the first instance, *then* recognise and critically respond to how it impedes one's understanding of the initially bewildering other. Gadamer points out that the historical etymology of 'prejudice' points to a non-pejorative connotation of provisional judgement, verdict, or decision that is valid prior to a final judgement (literally a pre-judgement).[43] Whilst tradition and its constitutive prejudices make it possible for us to have a perspective – to *see* in the first place – they nonetheless fluctuate, change, and are

[42] Dreyfus and Taylor, *Retrieving Realism*, p. 112.
[43] Gadamer, *Truth and Method*, p. 279.

open to challenge, making them a fruitful basis for the advancement of understanding. This point is tangible in the common law, for example, where precedent and prior judgments form the interpretive lens through which a case is understood and decided. Yet later decisions can utilise prior judgments in a different way, given changing societal expectations, norms, or their application to unique circumstances. On one hand, the expansion or contraction of our range of vision is contingent on how our belonging to a particular tradition frames our values and beliefs. On the other hand, we have an active role in shaping their future possibilities; we participate in and determine their development, where faith in authority does not replace critical reflection. Some prejudices will reflect pernicious, destructive beliefs, but their scrutiny occurs through transformative, challenging encounters with others rather than solitary reflection. Genuine dialogue will therefore enact 'the art of testing' where 'one does not try to argue the other person down but that one really considers the weight of the other's opinion'.[44]

Gadamer calls this process of dialogical enrichment a 'fusion of horizons'. We initially acknowledge otherness – of being addressed by another, or 'awakened to something'[45] – then reflexively challenge the validity of our prejudices in light of this otherness, and finally, overcome our initial experience of alienation so what *was* alien is viewed now as one of many possibilities. 'We make conscious the prejudices governing our own understanding so that . . . another's meaning, can be isolated and valued on its own'.[46] Rather than dismissing a view outright, 'one learns to look beyond what is close at hand – not in order to look away from it but to see it better, within a larger whole and in truer proportion'.[47] Gadamer further states that 'each person opens himself to the other, truly accepts his point of view as valid and transposes himself into the other to such extent that he understands not the particular individual but what he says'.[48] That understanding revolves around *what another says* as opposed to the particularity of another individual makes sense when we consider the recognitional bonds grounding the process of dialogical enrichment. We recognise the distinctiveness of others by acknowledging their own differing conceptual framework that informs their unique interpretive perspective, expressed through *what they communicate*. Claims that we

[44] Ibid., p. 360.
[45] Jean Grondin, *The Philosophy of Gadamer*, trans. Kathryn Plant (Chesham: Acumen, 2003), p. 97.
[46] Gadamer, *Truth and Method*, p. 298. [47] Ibid., p. 304. [48] Ibid., p. 387.

understand the individual suggest that we *mediate* the other, essentially robbing them of their separateness and independence.

The importance of recognitional dialogical practices is clearest where they are absent, resulting in the harms of *objectification* and *assimilation*. Objectification has two meanings in this context: first, objectification implies a particular truth-status: when we say something is objectively true, it means it captures how things really are. At a deeper level, this implies something about our epistemic stance, in the sense that we have adopted a disengaged interpretive stance *towards ourselves* so that our perspective is no longer consciously situated within a prejudicial horizon that is open to examination, challenge, or change. If one's standpoint is objectified, it can represent epistemic and dialogical closure towards the conflicting perspective of others. This disengagement becomes externalised further, leading to the second occurrence of objectification: the rejection of another's subjectivity. This happens when our understanding of others no longer revolves around them as subjects with their own views and perspective. When we treat others as 'objects', we view them instrumentally, as something to be used like a tool for our purposes – a mere means to promote our own ends.

Objectification consolidates implicit power dynamics that often exist simply by virtue of unequal psychological capacities and asymmetrical moral responsibilities. It uses power to shape narratives based on exclusionary prejudgements, characterising certain individuals or groups as cognitively incompetent, unreliable, or untrustworthy, for example. Individual with impairment are not a participants within dialogue, but relegated to the role of 'passive bystanders', incapable of exercising epistemic and practical agency.[49] Their identity becomes bound up with another's purpose, unworthy of independent moral consideration or toleration. They are, moreover, *problematic* individuals to be dealt with because not only are they seen to fail the evaluative standards set by one's unreflective prejudgements, but they lack the basic capacity to conform to such standards, even if they wished to.[50] On these grounds, their manipulation and use is both justifiable and necessary. As these narratives acquire the veneer of objectivity and truth, they become the normative standard by which others judge themselves, leading to what Hildemann Nelson calls 'infiltrated consciousness'.[51] The internalisation of these narratives

[49] Miranda Fricker, *Epistemic Injustice: Power and the Ethics of Knowing* (Oxford: Oxford University Press, 2007), p. 132.

[50] Lindemann Nelson, *Damaged Identities*, p. 173.

[51] Ibid., p. 107.

fundamentally damages another's practical identity. Numerous cases of abuse and neglect involving individuals with impairments reveal that this type of engagement is all too frequent. Examples include the systematic harassment and intimidation used to exercise control of the money or estate of individuals with impairments.[52] Or when behavioural cues of acute distress and pain are interpreted as 'acting out' as a result of one's impairment.[53] Or when individuals with learning disability are given a 'Do not resuscitate' order without their or their family's consultation, on the presumption that such lives are not worth preserving.[54] We saw in the previous chapter the intuitive wrongness of this sort of treatment and how third-party interventions can be sanctioned as a result.

The harm of dialogical *assimilation*, on the other hand, is more subtle: it occurs when the prejudicial horizon of another comes to dominate the individual with impairment so that '[o]ne claims to know the other's claim from his point of view and even to understand the other better than the other understands himself.'[55] One's perceptions and thoughts come to stand for those of the other, so that their claims are 'co-opted and pre-empted reflectively from the standpoint of the other person . . . to the point of the complete domination of one person by the other'.[56] Like objectification, assimilation represents dialogical closure in two ways: first, we remain unaware and unreflective of our prejudicial standpoint, given that our presumptions are taken to be true and representative of another. Second, understanding of another individual is foreclosed because they are assumed to be mere reflections of ourselves; they are no longer recognised as independent, separate beings from ourselves. Power and domination is wielded in more indirect ways compared to objectification. Narrative damage or the manipulative use of another may not be the intent – indeed, the motivation might even be perceived as benevolent, where persons responsible are unconscious of their misuse of power. Assimilation can nonetheless result in harms to another's practical identity. Their unique perspective is appropriated so as to function as a mere reflection of ourselves, thus invalidating self-constituting narratives that contradict

[52] *London Borough of Redbridge v G, C, and F* [2014] EWHC 485 (COP); *A Local Authority v DL & Ors* [2010] EWHC 2675 (Fam); *A Local Authority v DL & Ors* [2011] EWHC 1022 (Fam).

[53] Mencap, *Death by Indifference: Following Up the* Treat Me Right! *Report* (London: Mencap, 2007).

[54] Mencap, *Death by Indifference: 74 Deaths and Counting; A Progress Report 5 Years On* (London: Mencap, 2012), pp. 15–16.

[55] Gadamer, *Truth and Method*, p. 353. [56] Ibid.

the understanding that is imposed on them. In short, 'by claiming to know him, one robs his claims of their legitimacy'.[57]

The problem of assimilation illustrates acutely the delicate balance that is demanded in hermeneutic competence. Empathic attribution of meaning of some sort is presupposed in any attempt to understand another. Empathy or putting ourselves in another's position is a common impulse and can often be the first step towards attenuating our perception towards another so as to glean clues about her internal states and intentions. Family members and caregivers often attribute thinking to the care-recipient with severe impairment, and this remains an important aspect of accepting, enabling relationships.[58] A productive kind of empathy can be defined as *compassionate regard*, comprising both agent-neutral and agent-relative dimensions, in order to differentiate it from assimilation. Agent-neutral judgements focus on our common humanness, objectively appreciating our shared fragile relationship to goodness and flourishing, rather than distinct impairments of another individual.[59] This cultivates an engaged bond and sense of joint purpose. Expressions of vulnerability are perceived as non-pathological, comprehensible responses to particular events, where it is assumed that 'the person is in touch with his or her surroundings and is expressing human emotion in familiar, shared-in-common ways.'[60] This common bond makes us further aware of how another's condition or state can impact on our own sense of flourishing, so that 'the person who is understanding does not know and judge as one who stands apart and unaffected but rather he thinks along with the other from the perspective of a specific bond of belonging, as if he too were affected.'[61]

Yet compassionate regard also directs one's attention towards agent-relative appraisals of another's particular condition or state. Agent-relative appraisals focus on how a condition or state applies to *certain individuals* – in the sense that it recognises an individual's unique experiences, personality, or suffering.[62] We lose sight of the 'size' of another's suffering, for example, if we believe that the hand pain that prevents a violinist from playing is equivalent to the irritating splinter in one's hand. As Nussbaum suggests, 'awareness of one's separate life is quite important if empathy is to be for another, and not for oneself, that one feels compassion, one

[57] Ibid., p. 354.

[58] Robert Bogden and Steven J. Taylor, 'Relationships with Severely Disabled People: The Social Construction of Humanness', *Social Problems* 36 (1989), p. 139.

[59] Kong, 'The Space between Second-Personal Respect and Rational Care'.

[60] Bogden and Taylor, 'Relationships with Severely Disabled People', p. 142.

[61] Gadamer, *Truth and Method*, p. 320.

[62] Bogden and Taylor, 'Relationships with Severely Disabled People'.

must be aware both of the bad lot of the sufferer and of the fact that it is right now, not one's own'.[63] Unlike assimilation, compassionate regard must integrate an accurate assessment of the unique abilities and limitations of others as well as ourselves, effectively restricting the extent to which we can 'take the role of another'. When compassionate regard is absent, empathic attribution can easily collapse into assimilation. Bogden and Taylor cite an example of a foster mother who 'makes decisions about how to treat her foster daughter by pretending she is the daughter and experiencing her actions' and 'reported experiencing, vicariously, the pleasure of being taken care of by looking at what she is doing for her foster child from her perspective'.[64] This need not always be pernicious or indicative of missing recognitional bonds: when an individual has the reflexive awareness that, even as these empathic imaginings facilitate an engaged emotional connection, they are likely inaccurate projections of one's own prejudicial standpoint, the harms associated with assimilation could potentially be mitigated to some extent. But the problem occurs when it is assumed that these imaginings amount to understanding and knowledge of the other: in believing that we make the other 'intelligible', based on our own experiences, language, conceptual schemes, we end up distorting them, not seeing them in their own terms.

Numerous examples of enmeshed relationships show how easy it is to fall into this trap. For example, *A Local Authority v M & Ors* [2014][65] featured a dispute about the future care and residence of M, a 24-year-old man with autism and a learning disability. The excessively controlling care regime of M's mother in particular was said to undermine M's social development, his 'ability to have a voice', and make independent choices.[66] Baker J concluded M's parents 'have such clear ideas about all aspects of M's life and believe the conclusions they have reached about him are correct', making it difficult for them to accept views that may differ from them.[67] Assimilation likewise occurs all too often when the perspective of the individual with impairment goes unrecognised entirely. In judging the capacity of an elderly gentleman with dementia in *A, B, & C v X & Y* [2012][68], Hedley J criticised the disputing parties for disregarding the perspective of X, stating that although both sides sought his best interests,

[63] Martha Nussbaum, *Upheavals of Thought* (Cambridge: Cambridge University Press, 2003), p. 300.

[64] Bogden and Taylor, 'Relationships with Severely Disabled People', p. 140.

[65] *A Local Authority v M & Ors* [2014] EWCOP 33.

[66] Paras. 60 and 30.

[67] Para. 250.

[68] *A, B, & C v X & Y* [2012] EWHC 2400 (COP).

'no one appears to have understood how their conflict appears through his eyes. Moreover, so entrenched have they become that, sadly I do not think anyone has even tried seriously to do so, as opposed to simply assuming that he would react as they would have done were they in his position.'[69]

Contra objectification and assimilation, genuine dialogue is premised on recognitional bonds, where the unique personal characteristics, preferences, feelings, and motives of individuals with impairments are deemed worthy of consideration. Our dialogical orientation must be rooted in humility to establish these recognitional bonds. Humility refers to a 'grounded orientation', where we have a realistic measure of our abilities and the surrounding circumstances.[70] There are strong, intuitive reasons for its importance in asymmetrical relationships. As seen earlier, exploitation of power imbalances and unequal psychological abilities can be all too common. Humility redresses this imbalance in the first instance because we face and appreciate our limitations. Consider the phenomenological experience of unilateral responsibility for another, especially in relationships involving individuals with severe impairments. We may be faced with questions of what to do, what is appropriate, how to understand another – yet have no immediate answers. We encounter a visceral sense of our limitations and potential deficiencies. These scenarios naturally arouse a sense of our vulnerabilities – both in terms of the vulnerable nature of our embodiment, but also of the initial inscrutability of another individual, of their complex needs and unique abilities. Eva Feder Kittay illustrates this point in her affecting account of when her daughter, Sesha, was diagnosed with a severe impairment:

> Sesha would never live a normal life. It would be another year before we completed the tests, the evaluations, the questionings that confirmed those first predictions. We couldn't know or fully accept the extent of her impairment, but some things were clear. We knew it wasn't a degenerative disability and for that we were grateful. But the worst fear was that her handicap involved her intellectual faculties. We, her parents, were intellectuals. I was committed to a life of the mind. Nothing mattered to me as much as to be able to reason, to reflect, to understand. This was the air I breathed. How was I to raise a daughter that would have no part of this? If my life took its meaning from thought, what kind of meaning would her life have?[71]

[69] Para. 15.

[70] Joseph Kupfner, 'The Moral Perspective of Humility', *Pacific Philosophical Quarterly* 84 (2003): 249–69.

[71] Kittay, *Love's Labor*, p. 150.

We might struggle to meet another's needs initially when faced with another's dependency, particularly if our life takes meaning from certain things another won't be able to share and we encounter challenges to our very ability to empathise and understand. Humility helps turn around these initial obstacles to our empathy, making us aware that we might not have all the right answers. Indeed, strongly held beliefs become challenged, and we begin to see ourselves as flawed and limited. Not in a self-denigrating sense, but in a way that makes us appreciative and open to others so we improve ourselves and grow.[72] This rightly functions as an immediate check on our arrogant impulses to control and dominate another.

Humility likewise triggers the development of an outward gaze. This enables us to recognise otherness in the first instance – of how an individual stands separate from us. We become open to the complex, unique features of another. But humility also involves becoming adept at distancing ourselves from personal desires and private interests, testing our prejudicial horizon whilst growing in awareness of and open towards 'more universal points of view' that emerge through 'the viewpoints of possible others', thus 'grasp[ing] the distinctions within what is opened to it in this way'.[73] In other words, one moves away from focus on the self, towards an other-regarding, more agent-neutral attitude 'which measures the importance of things independently of their relationship to oneself', of value features of the world that transcend the self.[74] Interactions with individuals with impairments challenge one's way of interacting with the world, teaching and improving one's understanding of what is worth valuing and not. Kittay states, for example, '[w]e didn't yet realize how much [Sesha] would teach us, but we already knew that we had learned something. That which we believed we valued, what we – I – thought was at the center of humanity, the capacity for thought, for reason was not it, not it at all.'[75]

In this way, reciprocity can emerge even in conditions of asymmetrical responsibility. This might seem impossible initially: if reciprocal dialogical practices require mutuality and equality, how can they occur in

[72] Kupfer, 'The Moral Perspective of Humility'.

[73] Gadamer, *Truth and Method*, pp. 15, 16.

[74] Thomas Hill, Jr., *Autonomy and Self-Respect* (Cambridge: Cambridge University Press, 1991), p. 112. See also Kupfer, 'The Moral Perspective of Humility', Nancy Snow, 'Humility', *Journal of Value Inquiry* 29 (1995): 203–16.

[75] Kittay, *Love's Labor*, p. 150.

relationships where one individual bears greater responsibility due to the other's impairment? Implicit in this question is a narrow understanding of reciprocity as a form of contractual exchange where like is returned for like, or the terms of exchange are, at the very least, commensurable and mutually consented to. However, reciprocity can be established in a second, deeper sense without mechanisms of consent. 'Reciprocity-in-connection'[76] is where individuals with impairment are seen as having much to offer, giving back perhaps in a different 'currency'. This could be through acts of love, affection, companionship, social ties or indeed, a heightened understanding of objectively important values, such as a more profound understanding of impairment, of human vulnerability, of a broadening circle of concerns outside the self.[77]

Humility is equally important in our engagement with those *outside* the personal-supportive relational sphere since promoting the decisional capacity of those with impairments will often involve a broader circle support. Both spheres share concerns over the individual, either in terms of her capacity to realise her autonomy or her well-being, and humility will do much to ensure that dialogical openness guides the search for common answers. Unproductive conflict results otherwise: consider the case of M again, whose family's trenchant prejudices meant they 'had difficulty in accepting the views of others where they differ from their own', even as these professionals could potentially contribute positively to decisions with and about M.[78]

To summarise, recognitional dialogical practices expressive of hermeneutic competence strike a fine balance between dual poles of complete disengagement and appropriation, between knowledge and ignorance towards one's own prejudicial horizon and the perspectives of others. This balance helps mitigate challenges associated with establishing dialogical reciprocity within a relationship of asymmetrical responsibility and unequal psychological competencies. Many cases like M reiterate the importance of having a dialogical orientation where one is simultaneously open and critically reflective towards both one's own prejudicial framework, as well as divergent, unique views others, be it the individual with impairment or outside professional support.[79] Dialogical closure can also

[76] Ibid., p. 67.
[77] Bogden and Taylor, 'Relationships with Severely Disabled People', p. 144.
[78] *A Local Authority v M & Ors* [2014], paras. 250, 28.
[79] Cf. *SCC v LM & Ors* [2013] EWHC 1137 (COP).

compromise the intersubjective cultivation of crucial self-nurturing traits necessary for the achievement of autonomy competencies, as we will see in the next section.

IV. Nurturing relations-to-self

So far, I have suggested that hermeneutic competence demands phenomenological attunement to the spatio-temporal conditions of coping and autonomy skills, and engages a recognitional mode of dialogical interaction. A comportment of dialogical openness is necessary not just in the treatment of individuals with impairments, but also towards supportive professionals who contribute to their care and development. Thus far, my discussion has focused on articulating features of the relational framework which reflects a hermeneutically competent dialogical standpoint. This phenomenologically attuned, recognitional stance is ethically significant for a number of reasons. Firstly, it maintains an important balance between reciprocity, humility, and respect for alterity in relational conditions of asymmetrical responsibility and unequal capacities. Secondly, it challenges assumptions about the priority of certain perspectives in the social space of reasons: a subjective, first-personal perspective about one's identity is only half of the story, given that its constituents are developed and realised within relational and social contexts; a third-personal, agent-neutral perspective, whilst important in considering the legitimacy of duties of support or intervention, can nonetheless potentially engulf or invalidate the subjective standpoint. The recognitional core of hermeneutic competence affirms the centrality of the *intersubjective, second-personal* perspective in cultivating the subjective resources and skills important for autonomy. This intersubjective dimension is especially important in light of the fact that our abilities for self-disclosure and transparency are limited. Yet self-understanding emerges in dialogical encounters with another standpoint or perspective.

I discussed in Chapter 3 that autonomy competencies presuppose a particular relationship to ourselves. Self-constituting narratives of our own value and worth enable us to act with the view that we are individuals worthy of respect and consideration.[80] This points to a self-nurturing core of autonomy that is 'not a matter of a solitary ego reflecting on itself, but is the result of an ongoing *intersubjective* process, in which one's attitudes

[80] Benson, 'Feminist Intuitions and the Normative Substance of Autonomy', in Taylor, ed., p. 135.

toward oneself emerges in one's encounter with an other's attitude toward oneself.[81] As Lindemann Nelson states, '[w]ho we can be is often a matter of who others take us to be. . . . Your identity as a competent adult crucially depends on others' recognizing you as such'. She continues, '[j]ust as we construct self-constituting stories around the aspects of ourselves and our lives we care most about, so *others* construct identity-constituting stories around the aspects of ourselves and our lives *they* care most about'.[82] Our 'affectively laden self-conceptions' depend 'on the sustaining attitudes of others'.[83] In other words, interactions grounded in these recognitional bonds promote crucial forms of *intrapersonal recognition* – of nurturing relations-to-self or ways of engaging with the self. Through the 'communicative enabling of self-realization'[84] our dialogical partner begins to emulate this discourse as a way of reflecting on and understanding herself. There are three modes of recognition relevant to autonomous, nurturing relations-to-self: recognition of the dignity of individuals as persons in social and legal environments becomes internalised as *self-respect*; the trust and closeness of intimate relational bonds are mirrored as *self-trust*, and the solidarity and belonging characteristic of membership within recognised communities leads to *self-esteem*.[85]

First, self-respect is the subjective sense of desert and dignity, where a person discerns her own status. Self-respect characterises one who possesses personal authority and thereby warrants deliberative respect. Recognitional bonds based on respect function as a mirror so that individuals come to adopt a similar disposition towards themselves, strengthening their conviction that they are worthy of certain treatment from others, as an entity with their own physical integrity, possessing personal characteristics and valid reasons others ought to acknowledge.[86] Objectifying and assimilating dialogical practices are diametrically opposite to recognitional bonds of respect, since those modes of engagement deprive

[81] Anderson and Honneth, 'Autonomy, Vulnerability, Recognition, and Justice', p. 131.

[82] Lindemann Nelson, *Damaged Identities*, pp. 81–2.

[83] Anderson and Honneth, 'Autonomy, Vulnerability, Recognition, and Justice', pp. 130–1.

[84] Axel Honneth, *The Struggle for Recognition; The Moral Grammar of Social Conflicts*, trans. Joel Anderson (Cambridge: Polity, 1995), p. 72.

[85] I am not so committed to Anderson and Honneth's particular taxonomy of the various relationships that correspond to attitudes towards the self. Some claims are too strong – for instance, the notion of self-esteem as intimately connected to communal ties bound through solidarity and shared values seems to be a questionable claim that I do not view as necessary for the development of self-esteem.

[86] See Darwall, *Second-Person Standpoint*, Anderson and Honneth, 'Autonomy, Vulnerability, Recognition, and Justice'.

another of ways of relating to themselves as unique individuals whose voice warrants consideration. A person who internalises degrading, disrespectful treatment will lack self-respect, making dissent to arbitrary disregard for her wishes unlikely.[87]

Structural recognition – as in the *legal* recognition of rights and personhood – is important for cultivating self-respect, for we see how its absence effectively sanctions individuals' unequal status compared to others within the political community.[88] If individuals with impairments are not accorded similar rights of personhood as those without impairments, this effectively reflects a lack of recognition and jeopardises their self-respect. Whilst important, these structural features are nonetheless insufficient for self-respect. For example, a suffragette fighting for the vote might still possess self-respect even without legal recognition of her equal status vis-à-vis men. The absence of structural recognition would likely inflict harm on her personal identity, but the deliberative respect accorded to her in surrounding relationships with likeminded women and progressive family members might bolster her self-respect such that her self-conception is robust enough to withstand the absence of legal recognition. In other words, respectful recognitional bonds within intimate relationships can sometimes mitigate deficits in structural recognition, though these inconsistencies will accentuate the interpersonal burden for sustaining such respect. The conferral of respect for one's status and authority in the primary-supportive relational sphere is therefore prior to structural recognition, particularly for the internalisation and development of respectful ways of relating to oneself.

Second, self-trust involves a confident, open connection to the perceptual, conative aspects of agency. We develop faith in our perceptions, emotions, desires, and impulses – the preconceptual constituents of the absorbed coping – when others validate our expressions of it, recognising their legitimacy and value.[89] Self-trust means we have a sense of the general reliability of these conative resources to provide us with situational cues that consequently orient our practical reasoning and agency. Dialogical openness and cultivated humility is of particular importance in

[87] See David Sachs, 'How to Distinguish Self-Respect from Self-Esteem', *Philosophy and Public Affairs* 10:4 (1981): 354. Sachs claims that it is virtually impossible to conceptualise an individual who lacks self-respect – but this overstates the case.

[88] Honneth, *Struggle for Recognition*, p. 134.

[89] Anderson and Honneth, 'Autonomy, Vulnerability, Recognition, and Justice', pp. 133–4. Also Agich, *Autonomy and Long-Term Care*, p. 83.

recognitional bonds that cultivate another's self-trust. When one demonstrates a confident, reflexive stance towards one's own coping responses, others likewise mimic and internalise this so that they trust their conative responses as not just important situational responses to particular circumstances, but catalysts for deeper reflexivity. Recognitional bonds based on trust and dialogical openness reinforce a sense of safety and compassionate regard to accommodate one's interactions with the surrounding environment, as well as facilitate explorations of one's inner life. If our relationships lack this open and trustful orientation, our sense of ourselves often mirrors this, sliding into psychological rigidity or complete uncertainty over our coping responses, and thus accentuating the often opaque, unreflexive nature of these conative dimensions.

Finally, self-esteem is the internalisation of the interpersonal recognition that one's role, activities, or goals are valuable contributions; it is when one mirrors esteem of oneself and one's activities. Recognitional bonds based on esteem affirm that one is an important participant in shared 'patterns of living' and communal rituals.[90] This link between esteem of activities and a sense of belonging is important. Recognition that we have an irreplaceable role and contribute positively to the collective aims of a group reinforces our sense of value and belonging, cultivating solidarity with others within a defined social community. In primary-supportive relationships, this means that one is not seen simply as a son or a daughter, but 'my son' or 'my daughter' – a crucial member of one's family.[91] Even the law mentions the importance of belonging. Hedley HJ in *Re GM; FP v GM and a Health Board* [2011][92] decided in favour of an elderly gentlemen returning home to his family despite the fact that a care home was likely to meet his physical and medical needs better. Hedley HJ stated, '[t]here is . . . more life than that, there is fundamentally the emotional dimension, the importance of relationships, the importance of a sense of belonging in the place in which you are living, and the sense of belonging to a specific group in respect of which you are a particularly important person'.[93] GM at home 'matter[ed] in a sense that he could never matter in an institutional care setting'.[94]

These self-nurturing relations-to-self are crucial constituents of autonomy skills – they bear on one's decisional capacity and ability to make authentic decisions that accord with a certain way of valuing oneself.

[90] Bogden and Taylor, 'Relationships with Severely Disabled People'. [91] Ibid.
[92] *Re GM; FP v GM and a Health Board* [2011] EWHC 2778 (COP).
[93] Para. 21. [94] Para. 23.

Their underlying recognitional mechanisms implicate the broader interpersonal context in a fundamental way: *particular ways of understanding and relating to ourselves reflect whether or not we have experienced and internalised different modes of recognition.* I said in Chapter 3 that individuals' self-conception must be the point of entry for any medico-juridical evaluations of the disabling or enabling quality of their relationships. My discussion here reveals further reasons why. Clearly, not all self-conceptions are equal: there are few – if any – cases where utter self-loathing, self-degrading, or self-doubting narratives can be deemed either accurate or normatively acceptable. Their predominance would indicate the exploitation of the asymmetry within a relationship in order to undermine an individual's potential agentic skills. By contrast, hermeneutic competence fuels recognitional mechanisms that exclude disabling dialogical practices surrounding an individual, and by implication, limit the types of self-constituting narratives that can be deemed credible. Narratives that reflect constituents of self-nurturing – self-respect, self-trust, and self-esteem – will gesture towards the presence of enabling relational, dialogical conditions that are necessary for the promotion of another's autonomy competencies.

V. Hermeneutic Competence in Capacity Assessment

Much of what I have said about the need for hermeneutic competence amongst the primary-supportive relational sphere of individuals with impairment likewise applies to the truncated timeline of mental capacity adjudications. These conditions are capable not only of enhancing or detracting from decisional capacity, but of influencing how meaning is conferred onto an agent's actions or words, affecting *external* interpretations of capacity. Two differences from the primary-supportive context must nonetheless be taken into account in the normative practices of medico-juridical assessments. Firstly, capacity adjudications often accentuate the power inequality between assessor and individuals with impairment. Assessors possess the authority to make legally binding determinations about capacity, decide what evidence is credible or relevant, and impose decisions that impact directly on a person's life through best interests judgments. Secondly, assessments have a more investigative, epistemic function that orientates the dialogical encounter accordingly. The medico-juridical assessor is one step removed from an individual's daily interactions with a more 'inquisitorial' purpose.[95] Affection and connectedness is a necessary (though not sufficient) characteristic of enabling primary-supportive relationships, but constraints of capacity assessments make this depth of engagement unrealistic. Particularly when conducted by court-appointed experts, such assessments are often impersonal, transient meetings, meaning there is little time or potential for establishing more intimate recognitional bonds that mitigate power inequities.

Particular dangers emerge with these distinguishing features. Even with the best intentions, unequal power dynamics are often exacerbated in such encounters. Capacity frequently becomes an issue when an individual makes decisions *against* professional advice. Though legal judgments have sometimes decided in favour of a patient's capacity, the adjudication itself might nonetheless represent an undue burden on individuals with impairments, especially considering the sanctity of patient autonomy amongst those *without* impairments. Diagnoses of learning or cognitive impairments often determine whether a person's refusal of recommended treatment or care is respected or questioned. Well-meaning interventions of medico-juridical practitioners can become unjustifiable interference, as I discussed in the previous chapter. Moreover, dialogical closure is a constant risk given the intrinsically interpretive nature of capacity adjudications. Even as the law posits that unwise decisions are not indicative of incapacity, capacity assessors' beliefs and values can sometimes cause capacity to bleed into strongly paternalistic best interests judgments.

Medico-juridical practice needs to be attentive to these dangers. The dialogical skill and recognitional mechanisms constitutive of hermeneutic competence can help mitigate these risks, functioning as a guide to good capacity adjudications so that every attempt is made to understand an individual's conceptual scheme, to grasp their self-conception as situated within their immediate relationships. For this, medico-juridical assessors need to adopt a reflexive stance towards their own conceptual framework. The concept of prejudice has particular resonance here. Medico-juridical assessments are simultaneously backward- and forward-looking, applying past judgments to new contexts. Situational particularities orient the assessor's interpretation of existing laws and previous judgments; conversely, application of these interpretations shapes the situational context

[95] Baker J states in *Cheshire West and Chester Council v P & Or* [2011] EWHC 1330 (Fam): 'The processes of the Court of Protection are essentially inquisitorial rather than adversarial' [para. 52].

in specific ways, bringing together the law's meaning and normative function. In other words, an assessor's interpretation of the MCA's principles and past judgments is unfixed, undergoing constant revision through the process of particularistic application. It is this act of application that simultaneously helps challenge entrenched prejudgements whilst opening future possibilities for different interpretations of mental capacity law. Values *beyond* legal principles and judgments likewise inform medico-juridical applications of the law, such as the importance of autonomy competencies and the values of social integration and personal welfare, as we saw in Chapter 3.

The complex mix of these influences in capacity adjudication necessitates a hermeneutically competent comportment that practises reflexivity towards one's prejudicial standpoint. Two reasons for this come to mind. First, *justificatory consistency* is better ensured, thus avoiding the disabling contradictions in third-party interventions that were explored in the previous chapter; second, the danger of *overdetermination* is better avoided in capacity adjudications. The next section explains these in turn.

VI. Justificatory Inconsistency, Overdetermination, and Narrative Repair

The previous chapter discussed the contradictory discourse of disablement and presumed incapacity in third-party interventions as illustrated in the case of WMA, where there was best interests decision to remove him from the stultifying care of his mother against his wishes. However, I argued that the disabling language that justified the intervention was problematic. I suggested that capacity assessments, and indeed, any justification for intervening in relationships due to their suppression of individuals' autonomy skills, should likewise be aligned with the language of enablement and potential decisional capacity. This might seem a mere semantic claim or a point about the need for logical consistency. But it goes deeper than this.

As mentioned earlier, language is a repository for our prejudgements and reflects our value orientation. Just as language through dialogue has the potential to build bridges, it can also entrench hidden prejudices that should be scrutinised or even abandoned, depending on their coherence, usefulness, and whether they reflect recognitional practices. A non-reflexive stance towards one's prejudices can result in substantial harm, inflicting the sort of 'epistemic injustice' Fricker writes about, compromising trusting relations and directly affecting the individual's self-constitution.[96] Consider, for example, the process of 'steadying the mind'. The mind becomes 'settled' or 'steadied' through trusting dialogue so that we come to favour beliefs that have credence and stability within the space of reasons; it is when we can trust in the 'uptake' of others that we can answer truthfully about our interpretations of how the world is ordered. Self-constitution relies on this process of 'steadying the mind', for 'not only our beliefs and desires but also our opinions and value commitments settle themselves through social dialogue into more or less stable states, so an important dimension of our identity thereby takes shape'.[97] Such relations of epistemic trust can matter in finely balanced capacity adjudications, as we see in *Wandsworth Clinical Commissioning Group v IA* [2014].[98] IA was resistant to undergoing a capacity assessment and refused to interact with particular medical professionals. The judge observed the importance of trust in mitigating IA's frustration with his circumstances, helping him manage his behaviour, and ultimately concluded that IA had capacity when considering the assistance trusted officials provided in helping him make decisions.[99] Through the lens of epistemic justice, the ability of care professionals to engage in trustful dialogue provide individuals with impairments crucial opportunities to 'steady their mind' and make joint decisions about their care and treatment. Rather than seeing them as 'a diagnosis' with preconceived prejudices about their emotional and personality traits, such individuals are recognised as worthy of social dialogue and engaging in mutually trustful epistemic relations.

Forms of epistemic injustice occur, however, when an interlocutor's prejudices impact on the perceived credibility of their conversation partner, excluding them from trustful conversation. Parts of their social identity – which also partially constitute their personal identity – form reasons for distrusting their status as competent knower. This results in isolation, systematic marginalisation, and epistemic distrust, where a person is excluded from participation within the type of engagement that helps develop necessary components of her identity. In Fricker's words, 'when

[96] I find Fricker's analysis generally convincing aside from her rejection of inferentialism. I do not believe an account of epistemic virtue needs to be contrary to a commitment to inferentialism, and indeed, her account of epistemic justice can in fact be separated from her arguments surrounding the epistemology of testimony. See Sanford Goldberg, 'Comments on Miranda Fricker's *Epistemic Justice*', *Episteme* 7 (2010): 138–50.

[97] Fricker, *Epistemic Justice*, p. 53.

[98] *Wandsworth Clinical Commissioning Group v IA* [2014] EWCOP 990.

[99] Paras. 63, 80, 81, 84, 89.

this is the case, the injustice cuts him to the quick. Not only does it undermine him in a capacity (the capacity for knowledge) that is essential to his value as a human being, it does so on grounds that discriminate against him in respect of some essential feature of him as a social being.'[100]

To apply this more tangibly to my discussion of capacity assessment, we may well accept that impairment is a feature of unique personal identity. This can, on one hand, lead to empowering, respectful attitudes towards individuals with impairments, as seen in *Wye Valley NHS Trust v Mr B* [2015][101]. The case revolved around the best interests decision of Mr B, an elderly gentleman with schizophrenia who refused a life-saving operation to amputate his severely infected foot. There were questions as to how much weight should be accorded Mr B's religious sentiments – where Mr B described angelic voices that prescribed whether he should take his medication. The Trust submitted that these amounted to delusional beliefs connected to his mental illness and should be overridden. In contrast, Jackson J argued:

> In some cases, of which this is an example, the wishes and feelings, beliefs and values of a person with a mental illness can be of such long standing that they are an inextricable part of the person that he is. In this situation, I do not find it helpful to see the person as if he were a person in good health who has been afflicted by illness. It is more real and more respectful to recognise him for who he is: a person with his own intrinsic beliefs and values. It is no more meaningful to think of Mr B without his illnesses and idiosyncratic beliefs than it is to speak of an unmusical Mozart.[102]

But, on the other hand, there is the temptation to reduce individuals to their impairment, taking as our starting point prejudgements about diagnosis which then automatically diminish such individuals' competence as both knower and agent. They then become excluded from recognitional dialogical practices that enable them to stabilise their mind and develop nurturing relations-to-self. When capacity assessments or third-party interventions utilise the discourse of disablement – of the intuitive 'untrustworthiness' of one's testimony due to one's impairment – they essentially reflect epistemically unjust practices that can have consequences on how individuals with impairments see themselves. This helps frame Jackson J's challenge to the Trust's submission that the best interests

[100] Fricker, *Epistemic Justice*, p. 54.
[101] *Wye Valley NHS Trust v Mr B* [2015] EWCOP 60. [102] Ibid., para. 13.

decision should not accord significant weight to his wishes, feelings, and religious beliefs. He stated,

> [Mr B] is a proud man who sees no reason to prefer the views of others to his own. His religious beliefs are deeply meaningful to him and do not deserve to be described as delusions: they are his faith and they are an intrinsic part of who he is. I would not define Mr B by reference to his mental illness or his religious beliefs. Rather, his core quality is his *"fierce independence"*, and it is this that is now, as he sees it, under attack.[103]

The second, related harm associated with non-reflexivity towards prejudice is overdetermination: when capacity adjudicators are unaware of their prejudices, this can overdetermine the outcome in capacity assessment. Hidden prejudices, or rigid certainty about them, can unduly orientate capacity adjudications, where the standards of knowledge and deliberation are raised arbitrarily and unconsciously according to diagnosis, to the point that the testimony of individuals with impairments becomes invalid. Eldergill DJ in *Westminster City Council v Manuela Sykes* noted, for instance, that Sykes was 'recorded that she has a tendency to become defiant when [issues about her return home] are raised'. But he continued, '[t]his is logical and understandable because, unless one has a memory of the previous difficulties, the professional view must appear patronising and intrusive, and the problems made-up or grossly exaggerated'.[104] The outcome, more often than not, is a finding of incapacity, or leads to a disproportionately paternalistic best interests decision. These issues are pertinent in the case *LBL v RYJ & Anor* [2010],[105] for example, where Macur J highlighted 'the difficulty [which arose] from *the approach of others* to the expression of [RYJ's] wishes'.[106] Most notably the

[103] Ibid., para. 43. One might wonder how this case differs from scenarios like Anne's in Chapter 3. I do not think that the situation with Anne's anorexia nervosa and refusal of treatment necessarily will differ in substantial terms from that of Mr B. The point about Anne's case was to illustrate how self-abnegating narratives can become an intrinsic part of oneself and that one needs more substantial evaluative tools to determine where her autonomy or decision-making has gone awry. But that said, in no way am I committed to the claim that in all circumstances, individuals with egosyntonic but self-harming disorders, such as eating disorders, are not deserving of being treated in a similarly competent manner as Jackson J's judgment clearly demonstrates in the *Wye Valley* case. Indeed, I make this case in my paper, 'Beyond the Balancing Scales' and in fact critique another judgment made by Jackson J in his best interests decision to impose life-preserving treatment on a severely anorexic individual.

[104] *Westminster City Council v Manuela Sykes* [2014] EWHC B9 (COP) §7.

[105] *LBL v RYJ & Anor* [2010] EWHC 2665 (COP).

[106] Para. 63, emphasis added.

capacity assessor failed to assist in her decision-making, given that RYJ's views 'were effectively challenged by continual repetitive questioning to "confirm" the same.' Macur J concluded that 'in those circumstances it is unsurprising that any person without impairment may begin to doubt that which they said initially.'[107] No evidence supported claims of incapacity through '"suggestibility" beyond the norm of those continuously questioned on the same topic and their answer apparently not accepted'.[108]

The dangers of overdetermination are even more striking in two other cases. First, the controversial and widely reported 'sparkly' case *King's College Hospital NHS Foundation Trust v C and Anor* [2015][109] revolved around whether C possessed mental capacity to refuse life-preserving dialysis. C was noted to have a long history of 'impulsive and self-centred decision making without guilt or regret'; she sought to 'live life entirely and unapologetically on her own terms', where she 'placed a significant premium on youth and beauty' and having a 'sparkly' lifestyle.[110] Following a number of stressful events (breast cancer diagnosis and treatment, relationship breakdown, financial debt and troubles, arrest, and criminal charges), C made an unsuccessful attempt to commit suicide, resulting in acute injuries to her kidneys. C refused to undertake further dialysis even though the prognosis was good for recovering kidney function. Clearly, C's decision was counterintuitive. As Macdonald J stated, it 'may not accord with the view that many may take in the same circumstances, and indeed may horrify some'.[111]

The capacity assessments were mixed: two out of the three witnesses argued that C lacked capacity, due to a possible personality disorder that caused dysfunctional, rigid thinking with regards to the optimistic prognosis for her kidney condition. The issue of overdetermination is most noticeable in Professor P's capacity assessment. He believed that C exhibited a '"petulant" response to a lack to a timely recovery'[112] and that she could not use and weigh information because she

> demonstrated no ability to consider and weigh alternative futures, no ability to place herself in her daughters' shoes when considering the effect of her refusing treatment or to weigh the impact on them of her suicide and no ability in respect of her prognosis to accept anything other than the inaccurate view that the damage to her kidneys is irreversible and she could not survive without permanent dialysis.[113]

[107] Para. 50. [108] Para. 53.
[109] *King's College Hospital NHS Foundation Trust v C and Anor* [2015] EWCOP 80.
[110] Para. 8. [111] Para. 91. [112] Para. 49. [113] Para. 49.

This capacity assessment seems to hinge on certain presuppositions about the type of factors one ought to consider in making this decision, or the type of values one ought to have, notwithstanding any understanding of the subjective character and personality of the individual in question. Contrast this with one of C's daughters whose statement directly challenged the claim her mother lacked capacity:

> [M]y mother would never have wanted to live at all costs. Her reasons for trying to kill herself in September and for refusing dialysis now are strongly in keeping with both her personality and her long held values. Although they are not reasons that are easy to understand, I believe that they are not only fully thought through, but also entirely in keeping with both her (unusual) value system and her (unusual) personality. Her unwillingness to consider 'a life she would find tolerable' is not a sign that she lacks capacity; it is a sign that what she would consider tolerable is different from what others might. . . . 'Recovery' to her does not just relate to her kidney function, but to regaining her 'sparkle' (her expensive, material and looks-orientated social life) which she believes she is too old to regain. Again, the references in the notes to her talking about being 'sociable, hosting parties and going out with the girls' are fitting: to those who know her well, her entire identity has been built around being a self-described 'vivacious and sociable person who lives life to the full and enjoys having fun'.[114]

When comparing the views of Professor P and C's daughters, the issue of how certain implicit prejudices can overdetermine capacity assessments is apparent. From a clinician's point of view, C's dismissal of her prognosis could only be explicable through a causal link to impaired thinking, which indicated her inability to use and weigh. This ignored the testimony and evidence of those who knew C best. V (another of C's daughters) charged that the Trust had 'place[d] the test for capacity too high by demanding from C a closely reasoned "balanced, nuanced, used and weighed position" in respect of her medical prognosis.'[115] She and her sister repeatedly recognised that this decision was fully concordant with C's values, personality, and self-conception, regardless of how unappealing or painful they were to those around her. They had 'learnt to accept their mother for who she [was]: complexities, seeming contradictions, blind spots, self-centred and manipulative behaviour, excruciating honesty and all', where 'there [was] no point in trying to make C a "typical mother"'.[116] To his credit, Macdonald J situated his capacity adjudication around C's own self-conception and narrative, ultimately concluding that the presumption of C's mental capacity had not been displaced, citing

[114] Para. 63. [115] Para. 69. [116] Para. 12.

specifically that 'her rejection of her prognosis is the result of her having considered it and given it no weight as against other factors more important to her'.[117]

A second case echoes this danger of overdetermination but also shows how it can be mitigated through hermeneutic competence. *LBX v K, L, and M* [2013][118] was a long-running case concerning L's capacity to make decisions about his residence, care, and contact with his family. The capacity assessor again made substantive assumptions about the type of knowledge and information that was required for each decision. Though L could communicate whether he wished to see particular individuals, 'he was not able to understand relevant information with regard to the positive aspects of his relationship with his father and brother, the risks of seeing people, the emotional factors, the importance of family ties, how the other person might feel about his decision and the nature of friendship'.[119] When challenged the assessor admitted that 'if what has been termed the "lower threshold" of information relevant to the decision was taken he considered it would be more likely than not that L had the potential to achieve mental capacity. If his somewhat higher threshold [was] adopted, he said it was highly unlikely L would have the potential to achieve mental capacity and he did not recommend any further assessment.'[120]

By contrast, the best interests assessment carried out by an independent social worker warrants deeper analysis as an admirable exemplar of hermeneutic competence. The social worker believed L 'more likely than not' had the potential to achieve mental capacity,[121] so long as others recognised his unique needs, modes of expression, and individuality:

> [F]irstly, what is considered to constitute relevant information by the person responsible for assessing L's mental capacity and whether that can be conveyed to L in a form that is tangible to him, to his tangible understanding of the realistic accommodation options available to him; secondly, that L is given all practical and appropriate support to enable him to make the decisions for himself; thirdly, having relevant information about the decisions provided in a way that [is] most appropriate to enable him to understand; and, fourthly, having the relevant information presented in a way that is appropriate for L's need and circumstances and explained using the most effective, tangible form of communication.[122]

The actual dialogical encounter between L and the social worker contained several notable features of hermeneutic competence. First is

[117] Para. 87. [118] *LBX v K, L, and M* [2013] EWHC 3230 (Fam) (2013) MLO 148.
[119] Para. 18. [120] Para. 21. [121] Para. 5. [122] Para. 31.

critical awareness of the ways in which one's prejudgements can shape dialogical engagement. The social worker 'went to great lengths to guard against [L's inconsistent answers to different people] by not using leading questions, by neutral responses in order to validate the process she undertook'.[123] A stance of dialogical openness was emphasised instead, as she stated explicitly the importance of 'ascertain[ing] L's understanding rather than impos[ing] any views about them on him'.[124] Second is the establishment of a communicative bond to achieve 'a fusion of horizons'. The social worker made significant efforts to establish a shared level of understanding through the use of communicative mechanisms. Drawings helped ascertain L's own level of understanding – whether he could distinguish between living at or visiting a place – as well as his views and preferences – whether he could place a 'happy', 'okay', or 'sad' card, as well as 'week' or 'weekend' cards, on pictures of each future accommodation. A similar procedure was carried out to determine L's wishes and feelings regarding contact with his family. Throughout this process she

> endeavoured to enable and encourage L to participate in the assessment as fully as possible by paying regard to the SALT [speech and language therapy] advice regarding tips to aid L's communication and understanding: use of short simple sentences, ensuring to speak at a normal volume with slow speed and basic words, taking extra time to pause and check understanding, breaking down difficult information into smaller points, allowing L time to understand and consider each point before continuing, repeating information where appropriate and summarising to check shared information.[125]

Finally, the social worker's best interest assessment took the individual's self-conception as the point of entry to determine whether his primary-supportive relationships sufficiently promoted his autonomy competencies. She noted that L was able to 'give reasoning for his expressed wishes and feelings . . . namely, wanting to stop contact with K [his father] and D [his brother] due to feeling pressure; an ability to evaluate the impact of pressure from D is different to that associated with M [his aunt]; being told to do naughty, silly things and having an ability to explain this; an ability to explain how this made him feel worry or to tell staff; an ability to weigh up some strategies for managing his predicament, i.e. "Even with staff, I worry about it".'[126] The social worker concluded that L was 'very able to communicate', with 'clear and understandable'

[123] Para. 33. [124] Para. 33. [125] Para. 28. [126] Para. 30.

expressive speech, whilst his 'actions to questions were coherent and the content was very consistent and congruent throughout.'[127]

These cases show that medico-juridical professionals – those charged with both capacity and best interests assessments – must display hermeneutic competence even as intimate connective bonds may be absent; they must exercise critical reflexivity towards their conceptual frameworks to ensure that recognitional, open dialogical practices are followed. As with the immediate primary-supportive relationships around individuals, capacity assessors must be open to engaging with and learning from another's point of view. This is important if we are to lessen the danger of overdetermination in adjudicating mental capacity. Whilst awareness of the impairment is important, it must not and should not determine the outcome of the dialogical encounter, otherwise it runs the risk of committing epistemic injustice against the individual. Capacity adjudications must guard against such forms of injustice; at the same time, such assessments are uniquely placed to identify the need for reparative mechanisms when disabling practices occur. Damage to an individual's capacity can be rectified through active resistance. However, they may lack the kind of community that can support such resistance, especially 'since the formation of such a community is itself a social achievement and not a social given.'[128]

In these scenarios, capacity assessments can recommend the need for narrative repair in the absence of supportive communities of resistance. Narrative repair seeks to overturn damage to practical identity; it is a means of resistance towards disabling ways of understanding oneself and reparation through recognitional 'counter-stories' that promote nurturing relations-to-self. These counter-stories seek to establish and encourage enablement, competence, inclusion, and moral worthiness, so that one can exercise crucial agential and autonomy skills with greater assurance. Counter-stories likewise try to shift others' understanding of the individual: though broader cultural change might not be imminent, this process will nonetheless initiate the expansion and enablement of one's autonomous practical agency. On one hand, capacity assessors have limited scope to carry out narrative repair. Their interactions with the individual are often not part of their daily interactions, and they are thus unable to practically implement the course of action necessary for this reparative process. On the other hand, however, these judgments are an important check on those who *can* enact change. Narrative repair seeks to cause

[127] Para. 27. [128] Fricker, *Epistemic Justice*, p. 54.

a shift in understanding from both first-personal and second-personal perspectives.[129] Capacity assessments function as significant opportunities not just to identify when narrative repair is necessary and appropriate, but also to *initiate* that process. They can affirm and reiterate the basic contours of reparative, enabling narratives. In sum, medico-juridical interventions – even through capacity adjudications – can precipitate change and, through the consistent application of enabling language based on recognitional, dialogical practices, provide the basic normative framework for the process of narrative repair in the first instance.

Conclusion

The temptation to focus solely on the competencies of individuals with impairments is strong, but it can shift our ethical attention away from the duties, responsibilities, and competencies that are necessary for those surrounding them. I have tried to argue for an ethical reorientation, so that the emphasis is on the interpersonal skills and abilities constitutive of a more relationally situated concept of capacity. Asymmetry can and often does exist in the relationships involving those with impairments – moral responsibility may be unilateral, respective capacities might differ. But this should not determine the mode of engagement that is owed to such individuals. Enabling relational practices acknowledge the asymmetry but mitigate its potential risks with hermeneutic competence – a holistic, interpretive orientation that fundamentally seeks to enhance the decisional capacity of individuals with impairments. We need phenomenological attunement to the embodied nature of absorbed coping and everyday autonomy if we are to share in the perceptual experiences of others; we need dialogical conditions premised on self-reflexivity, humility, and openness if we are to understand others without objectifying or assimilating them; we need recognitional practices that cultivate nurturing relations-to-self if we are to lay the groundwork for an individual's further autonomy skills and competencies.

This applies to both the *internal practices* and the *external interpretation* of capacity. The same dialogical practices regulating an autonomy-enhancing relationship should likewise describe the appropriate ethical-dialectical stance of the medico-juridical professional assessing capacity, without which interventions in disabling relationships could lack justification, commit epistemic injustice, or miss opportunities for narrative

[129] Lindemann Nelson, *Damaged Identities*, p. 156.

repair, even if only to provide its basic contours. What I have shown here is that medico-juridical assessments of capacity are themselves ethical-dialogical interventions that are situated within a specific prejudicial horizon, particularly when viewed from the perspective of the person whose capacity is in question. As the concluding chapter explores, this will raise a number of ethical implications for the practice of capacity adjudication.

7

Rethinking Capacity

This book has argued that the context of impairment demands a *relational* concept of mental capacity. Capacity draws upon a range of implicit values and goods; chief amongst them is the importance of perceptual and autonomy competencies constitutive of temporally extended agency, all of which are promoted through surrounding relationships and supportive environments. Capacity understood as such results in a profound ethical reorientation, where the burden of practical competency shifts away from the individual in question, towards the ethical-dialogical stance of her surrounding relationships, of the capacity assessor or other intervening professionals. We might see this as a radical departure from current medico-juridical practice, but what is ethically at stake in certain legal cases can be captured more accurately through this relational lens of capacity. I want to use this final chapter to briefly summarise five key implications of my argument, in terms of what this means for a person's primary relationships as well as for the future of capacity adjudication.

1. Autonomy must incorporate phenomenological, relational dimensions

Autonomy is an important value in the medico-juridical sphere. It isn't the only value we should prioritise. But given that mental capacity is often used as the shorthand threshold concept for the respect of personal autonomy, it is crucial that we get our operating concept right, ensuring its attunement to the realities of impairment. There is a tendency to privilege the rational, the cognitive, the moments of making important, deliberative choices. But this obscures the grounding role of our perceptual skills, of our embodied ways of having contact and engaging with the world. The universal, outward, and intercorporeal nature of absorbed coping illustrates, at a very basic level, how our bodies seek connection with the world and become socially inscribed with meaning as a result. Like its perceptual counterpart, the socially acquired nature of autonomy competencies – skills that help us reflect on our subjective preferences and values – make us vulnerable to our relationships and social environment. The latter helps shape who we are, how we relate to ourselves and others, how we understand our abilities and frame the possibilities in our lives. *Connection* is often necessary for autonomy to flourish – our bodily connection with our physical environment, with spatial and temporal dimensions of experience, and our emotional, intellectual bonds with others, ranging from our primary relationships to those who support us in a professional role to legal recognition by the state. Impairment can present perceptual challenges that manifest themselves in particular ways of absorbed coping and can likewise generate greater vulnerabilities to surrounding narratives. This needs to be properly appreciated if the autonomy of individuals with impairments is to be fully developed and respected.

2. The competence of those around individuals with impairments matters

Too often the primary focus revolves around the particular abilities of the person whose mental capacity is in question. But such individuals do not exist in a vacuum – none of us do. Enabling relationships inculcate and promote self-nurturing narratives and ways of understanding the self in individuals with impairments. Relationships demonstrating hermeneutic competence will transcend the dichotomy of respect and care. Varying degrees of vulnerability, asymmetrical abilities, and responsibilities may describe the basic structure of these relationships, but they do not determine the mode of engagement that is ethically owed to individuals. It is all too easy to make the impairment define the person, to close ourselves from truly understanding her perspective and conceptual framework. And when this occurs, the dangers of objectification and assimilation intensify. Those with hermeneutic competence acknowledge the presence of impairment but do not make this constitutive of the individual. They will recognise certain intrinsic difficulties but mitigate them through phenomenological attunement and recognitional dialogical practices, seeking to establish shared perceptual experiences, recognising the intrinsic separateness of persons, and attempting to understand their unique perspective, voice, and personality, to work towards social belonging and temporally extended agency.

3. The content of self-constituting narratives matters

Regardless of whether we have an impairment, we are vulnerable to the narratives around us: if a child is constantly told she is useless, helpless, unworthy of consideration, it will have a profound impact on how she sees herself in future. When internalised, these potent narratives become part of her.[1] Her future agency – her sense of her abilities and subsequent choices – is marked accordingly. The unique perceptual and cognitive challenges one faces depending on the unique nature of one's impairment makes the surrounding types of intersubjective narratives and modes of interpersonal engagement even more important. Diminishing, denigrating narratives – those that see the 'diagnosis' before the unique person, minimise individuals' self-expression through control, abuse, or manipulation, or reinforce relational enmeshment and extreme helplessness – can never ground the capacity and autonomy of individuals with impairments. On one hand, surrounding others must be attentive to the phenomenological aspects of absorbed coping and be attuned to how impairments may disrupt perceptual and reflective skills. On the other hand, recognitional, self-nurturing narratives help develop another's self-esteem, self-trust, and self-confidence in appropriate ways so their agency can be realised.

In sum, relationships that enable and promote individuals' decisional capacity will have the following features:

1. They recognise that autonomy is not coextensive with rugged independence, that relational support is a crucial constituent that helps individuals realise their potential decisional abilities.
2. They establish a shared perceptual world through understanding and accommodate how impairment may impact one's expression of embodied coping.
3. They display sensitivity and attunement to the vulnerabilities unique to one's impairment, even as they cultivate the constituent skills of autonomy.

[1] Quite literally, as research shows how childhood stress and trauma can lead to epigenetic changes within the brain. See Donna Jackson Nakazawa, *Childhood Disrupted: How Your Biography Becomes Your Biology and How You Can Heal* (New York: Atria, 2015).

4. They recognise the separateness of persons, accepting them as they are, that they still have their own unique perspectives and views regardless of impairment.
5. They are conscious of, and actively guard against, harms of objectification and assimilation, by maintaining an engaged stance of dialogical openness, humility, and conscious awareness of their own interpretive limitations.
6. They are open to the ways in which individuals may express reciprocity even as asymmetrical moral responsibility may characterise the basic structure of the relationship.

4. The boundary between capacity and best interests is blurry and elastic

Current legislation suggests that capacity is a crude either-or scenario: either individuals have capacity (and therefore have the right of autonomy respected) or they do not (and therefore can be forced to accept paternalistic decisions made on their behalf). A bright line separates the legal boundary between capacity and best interests. By implication, the type of judgements one is permitted to make is determined accordingly. This is simply an empirical fact in one sense: the outcome of capacity assessments *does* result in an either-or decision: either we intervene or don't in an abusive environment; either we force life-preserving treatment onto an unwilling individual or we respect her refusal, eventually leading to her death. But at a deeper level, this way of conceptualising the boundary between capacity and best interests – though understandably concrete for the purposes of legislation – is neither phenomenologically accurate nor merited from a philosophical perspective. Like autonomy, capacity – and what we owe to individuals *even* when they lack capacity – must be conceptualised as a spectrum.

We might think this means the following: capacity can be *quantified* on a comparative scale, so that some may have *more or less* capacity and their right to have their subjective choices respected increases or lessens accordingly. If the legal threshold of mental capacity represents the number 5 (with 10 being 'full' capacity), the closer an individual is to 5 (i.e. 4 or 4.5), the more participation and weight will be accorded to her wishes in the best interests decision. Conversely, the further away a person is from the legal threshold, the less participatory and determinative her wishes and preferences will be in deciding her best interests. In that case, we might argue that even though *capacity* is invariably a black-and-white judgment, the law already thinks of her *right of autonomy* as a spectrum, exactly in this manner.[2]

This only partially captures the implication that follows from my book. I take it that the actual outcome of medico-juridical capacity assessment always appears like a stark cliff-edge. And whilst it may be the case that the right of autonomy can be teased apart from capacity, a rather narrow understanding of what that right entails is presumed, that is, that we defer to an individual's subjective wishes. This neglects two key points implied in a spectrum view of capacity and best interests: first, capacity assessments deploy the same types of substantive judgements that are deemed permissible in best interests decision-making, though the strict boundary between capacity and best interests suggests otherwise (i.e. that the capacity assessment remains devoid of passing judgement about the content of the individual's decision or the context in which she makes that decision). Rather tellingly, capacity adjudications appear from one direction to increasingly bleed into (implicit) best interests judgments, yet from another direction, individuals' participation, values, and rights figure more prominently in recent best interests decisions. This might be because judges are implicitly quantifying a person's capacity in the manner suggested above. But I suspect this has more to do with the fact that the value judgements typical of best interests considerations inevitably seep into capacity adjudications, and indeed, vice versa, particularly as the practical and indeed, normative, limitations of the cliff-edge view of capacity become apparent. In *Wye Valley NHS Trust v Mr B* [2015], Jackson J rightly states:

> As the Act [MCA] and the European Convention make clear, a conclusion that a person lacks decision-making capacity is not an *"off-switch"* for his rights and freedoms. To state the obvious, the wishes and feelings, beliefs and values of people with a mental disability are as important to them as they are to anyone else, and may even be more important. It would therefore be wrong in principle to apply any automatic discount to their point of view.[3]

Jackson J doesn't say so, but the shift to the language of universal legal capacity in the Convention for the Rights of Persons with Disabilities may reinforce precisely the same point about the valid consideration of

[2] *ITW v Z & Ors* [2009] EWHC 2525 (Fam); *A Primary Care Trust v P & Ors* [2009] EW Misc 10 (EWCOP).
[3] *Wye Valley NHS Trust v Mr B* [2015] EWCOP 60, para. 11.

individuals' rights, wishes, and values, *regardless* of where they fall within the spectrum of mental ability.[4]

This might bode well for improving best interests judgments, but what about capacity assessments? The flip side, one might worry, is that capacity adjudications become inherently disrespectful, where findings of incapacity are conflated with substantive judgements about the merit of one's values and choices. In one sense (and as I have discussed throughout), I think the infiltration of value into capacity assessments is inevitable; it is false, and indeed misleading, to presume otherwise. Capacity assessments must nonetheless be explicitly reflective and aware of the underlying concepts and values at stake, so that proper democratic debate can be had about how capacity assessments are undertaken.

This leads to the second implication of a spectrum view: if we start to consider individuals within context – from their subjective impairment and self-narratives, to their relational environment – then adherence to a stark distinction between capacity and best interests becomes harder, given that appropriate interventions will often be based on trying to promote enabling conditions to help develop individuals' autonomy skills. Capacity is not the arbitrary cutoff point from which we assist individuals or not: as this book has shown, those who technically have 'capacity' often consent to relationships that neglect, abuse, manipulate, mistreat them. The boundary between capacity and best interests needs to be sufficiently elastic so as to allow greater, more justifiable scope for positive, beneficent interventions in disabling circumstances, even against a person's wishes. Ultimately, a finding of capacity should not sanction neglect.

How this would look in practice would still need to be developed. However, it seems to me that a more elastic boundary between capacity and best interests would have the following preliminary features, of which some judgments already enact in practice: (i) reference to values would not be confined to best interests judgments, but would be openly discussed and debated in capacity adjudications; (ii) individuals would be subject to fundamentally enabling capacity assessments and best interests judgments, premised on self-nurturing recognitional mechanisms, regardless of where they fall in the spectrum of mental capacity; (iii) there would be justifiable (but carefully limited) scope for intervening on consented-to abusive relationships that impede individuals' mental capacity.

[4] I distinguish this from the 'will and preferences' paradigm examined in Chapter 2.

5. Capacity as a socially situated, relational, and dialogical concept transforms the role of the capacity assessor

Capacity adjudications often have an impartial, objective quality, where assessors allegedly adopt an agent-neutral stance towards the decisions and values of the individual. Clinicians are actively encouraged to recognise their own professional values when engaging in a therapeutic dialogue with patients to determine their decisional capacity, yet this is notably absent in the law. As Donnelly notes,

> Ultimately, and inevitably, all capacity assessors come to the task clothed with their professional and personal values, motivations and beliefs. These factors impact on how assessors engage with the people whose capacity they assess and may determine the conclusions they reach. Yet for the most part, the law operates as if these factors did not exist.[5]

Often legal judgments of mental capacity are delivered in a manner that suggests the judge is an impartial observer who stands outside the primary-supportive relationships of the individual being assessed. In one superficial sense, this is true: the judge is not personally involved with the patient and her relationships. Further, the judge is called upon to arbitrate conflicting views about a person's capacity and decide based on the evidence that both sides present. Probing deeper, however, we have seen that medico-juridical assessors are embedded within their own space of reasons and engage in an interpretive task when deciding about an individual's capacity. The interpretive task activates their own prejudgements – about rationality, about the nature of legal rights, about how autonomy should be understood, about the values that are worthwhile and frame the law.

A key implication of my argument is a fundamental rethink about the normative role and function of the capacity assessor. Part of the task of this book has been to render explicit how the medico-juridical assessor is dialogically situated in capacity assessments: to articulate how implicit assumptions about practical deliberation, autonomy, and relational support help determine capacity adjudications. The law is a crucial framework of meaning, with a range of presuppositions and values that can have profound consequences for individuals with impairments. It is vital that these presuppositions become explicit if more fruitful dialogue about decisional capacity is to be had. When autonomy as a value is invoked, for example, it is not the case that one 'standard' view of

[5] Donnelly, 'Capacity Assessment under the Mental Capacity Act', p. 480.

autonomy is correct; indeed, oftentimes practitioners utilise autonomy in incommensurable ways within the same case: one person might be advocating for autonomy in the sense of 'deferring to a person's wishes', whilst others might be referring to autonomy in the sense of 'promoting this individual's abilities and capacities'. When capacity assessors say, 'this person cannot use and weigh', what embedded presumptions are they making about rationality and reasoning? This book has sought to clarify some of these resonant values and concepts that operate both explicitly and implicitly in medico-juridical assessments, loosening assumptions about value- and content-neutrality that may prevent medico-juridical practitioners from debating openly, particularly in capacity adjudications (as opposed to best interests judgments). To arbitrate between incommensurable values and reasons, one often has to fall back on one's judgements that take into consideration other, often extra-legal factors – the condition of care, the individual's self-conception, the specific relational context, the type of knowledge one thinks is relevant – or the hypergoods that move us – the primary value of life, the priority of individual choice, the importance of human dignity, the good of social inclusion.

In this sense, assessors must work towards even greater transparency in their capacity adjudications, recognising how these judgments disclose and articulate certain values. There needs to be greater recognition of the fact that capacity adjudications are themselves forms of intervention, welcome or not, into the lives of individuals with impairments, even if they end up deferring to the person's subjective preferences. There is no getting away from the at times intrusive nature of such assessments.

To mitigate this feature, hermeneutic competence is as applicable in these circumstances as it is in individuals' primary-supportive relationships. Capacity assessors in the medico-juridical sphere – including judges who adjudicate evidence – must recognise their power vis-à-vis the individuals in question. In practice, judges and legal professionals often don't meet the person whose capacity is debated, taking as their point of departure the medical testimony or evidence presented to them. But if capacity is a dialogical concept, it seems to me that *ethically driven* capacity adjudications will demand first-hand dialogue with the individual herself. For example, consider how we intuitively, automatically fill in the gaps about someone, based on our own perspective, when we engage in written correspondence with someone we've never met. I might interpret a short sentence as excessively terse or someone repeatedly misspelling my name as inconsiderate. What is written can be easily misinterpreted. But as soon as I meet and converse with the person in question, these perceptions of mine might be proven to be completely wrong. The individual with no truck for extra written niceties might come across as extremely warm personally, or the repeated mis-speller of names generous and thoughtful. The point is that once we speak to someone in person, we get a better sense of what she's about, where she's coming from, what she means. We have greater opportunities to improve our understanding of the other person.

The same interpretive pitfalls of written correspondence apply even more so to capacity adjudications where those charged with deciding capacity don't actually meet the person in question. According to my account, third parties are obliged to intervene in cases where individuals' autonomy competencies are suppressed or systematically neglected, but they must start with individuals' narratives and self-understanding to avoid these duties becoming overly paternalistic or perfectionist. How is it possible to understand and gauge an individual's self-conception, and the role relationships have in that self-conception, if those determining capacity don't even have a conversation with the person?[6] This applies especially in finely balanced cases. On one hand, practical and logistical constraints are not to be dismissed. Often legal professionals have extremely heavy caseloads, in which case, even if a conversation was to be had, it could likely be brief and rushed. But aside from these constraints, we need to consider the *ethical-normative* demands of a more relational concept of capacity, of how we can ensure that capacity assessments themselves exhibit hermeneutic competence, especially considering the intrusive nature of these capacity adjudications in the first place. Legal professionals charged with determining capacity in the courts often meet care and medical professionals, representatives from the local authority, family members, yet a vital half of the conversation is missing. First-hand dialogue with the individual in question ought to be required because it presents a key opportunity to understand the person, her standpoint and choices. It helps arbiters of mental capacity recognise the subjective impact of her decision: that these *are* interventions into people's lives regardless, making the same phenomenological attunement, dialogical openness, and recognitional practices that are normative for their primary-supportive relational context equally important.

[6] See, for example, the importance of KK's first-hand testimony in the judge's decision of *KK v STCC* [2012] EWHC 2136.

There has been little attempt to standardise the practice of judges meeting with persons whose capacity or best interests are to be determined.[7] But where this does occur, the decisions have been notably sensitive and attentive to the individual. Jackson J's judgment in *Wye Valley NHS Trust v Mr B* is an exemplar in this respect. He rejected the Trust's application to force a life-preserving amputation onto Mr B, having met him personally, stating:

> [G]iven the momentous consequences of the decision either way, I did not feel able to reach a conclusion without meeting Mr B myself. There were two excellent recent reports of discussions with him, but there is no substitute for a face-to-face meeting where the patient would like it to happen. The advantages can be considerable, and proved so in this case. In the first place, I obtained a deeper understanding of Mr B's personality and view of the world, supplementing and illuminating the earlier reports. Secondly, Mr B seemed glad to have the opportunity to get his point of view across. To whatever small degree, the meeting may have helped him to understand something of the process and to make sense of whatever decision was then made. Thirdly, the nurses were pleased that Mr B was going to have the fullest opportunity to get his point across. A case like this is difficult for the nursing staff in particular and I hope that the fact that Mr B has been as fully involved as possible will make it easier for them to care for him at what will undoubtedly be a difficult time.[8]

Adjudications at the legal level are often the top of the chain in disputes about capacity or best interests. This particular hierarchy presents additional ethical burdens and responsibilities onto practitioners of the law as a result. They are in a unique position to judge when third-party interventions are appropriate in cases where individuals' relationships or immediate care environment suppress and neglect their autonomy competencies. But even as they may be external to these primary-supportive relationships, they are not excluded from the dialogical norms that ought to structure enabling interactions. They must likewise demonstrate the types of enabling relations that are to be instituted further down the line, in the day-to-day treatment of individuals with

[7] Aside from Charles J's document, 'New guidance issued on facilitating participation of 'P' and vulnerable persons in Court of Protection proceedings' published in November 2016, available online: http://www.familylaw.co.uk/system/froala_assets/documents/1245/Practice_Guidance_Vulnerable_Persons.pdf (accessed 15 November 2016). Charles J emphasises the communicative and practical steps which need to maximise the involvement of P but he still ultimately defers to judicial discretion when it comes to decisions about judges meeting P in Court of Protection proceedings.

[8] *Wye Valley NHS Trust v Mr B* [2015] EWCOP 60, para. 18.

impairments, as Jackson J himself notes. When a legal professional meets with the person, it is a chance for initiating a process narrative repair – for bringing about a shift in understanding about the individual, directly to the individual and the primary-supportive relationships around her, thus setting a new tone that is normative for future interpersonal engagement.

Lessons from Ava and Rett Syndrome

Finally, I want to come full circle and conclude with a story more close to home. What I have written here will have little value if it has no traction with the real experiences of impairment. I mentioned earlier that my niece, Ava, has been diagnosed with Rett syndrome, a rare neurological and developmental disorder that affects girls almost exclusively. Rett syndrome is often described as a 'thief', in that a child's development appears normal for a time but is followed by a rapid decline in mobility, verbalisation, and purposeful movement. Needless to say, the eventual diagnosis was devastating, particularly for my sister and her husband, who grieved the future hopes they envisaged for her and their family.

Ava's impairments present substantial daily challenges: her ability to walk is declining, and she will eventually need to use a wheelchair. She has lost the use of her hands and has difficulty gripping things. She cannot speak or communicate with language. Her ability to eat is deteriorating, making a feeding G-tube likely in the future. She is unable to control her bowels and needs help cleaning and feeding herself. She has dystonia, seizures, bruxism (teeth grinding), abnormal muscle tone, autonomic dysfunction, and struggles to breathe and sleep.

Ava has what my sister calls a 'village' of eighteen therapists and volunteers to deal with these physical impairments: she goes to hydrotherapy and equestrian therapy, both of which Ava loves and are crucial to ensure her muscles don't atrophy. She wears ankle foot orthoses to help with her stability. Ava has an occupational therapist to help with daily living, such as feeding (as she struggles to gain weight and maintain nutrition) and is in the process of being trip-trained to use the bathroom in hopes that it will ease her transition into school. Physical therapists help her with stretching and exercises to try and delay the onset of scoliosis, a common occurrence in girls with Rett syndrome.

Crucially, these impairments have not stopped Ava from expressing herself to those around her. Ava communicates primarily with her eyes – she uses an eye gaze board, and getting an eye gaze computer in future will be crucial in helping her have a voice. In my sister's poignant words, 'Her eyes are really the window to her soul. She has the most intense eye gaze of anyone I know. It is the primary reason I believe her understanding is there but she is unable to get her body to cooperate.' Even with her limited hand use, she can gesture, bat, and swipe at things she wants. She expresses her dislike for certain foods by blocking it from entering her mouth. She shows emotions and will let you know when she wants something, is hurt, or finds something enjoyable or disagreeable. My sister cites this example: her two sons enjoy listening to a particular pop radio station in the car, which she will change to talk radio when they are not present. Ava will cry and make loud vocalisations until my sister changes back to the station, and sometimes she will immediately settle. She expresses joy when she does the things she loves: like swimming, equestrian therapy, watching *Dora the Explorer* and *My Little Pony*, playing with balls, being on swings, touching sensory toys. She can pick out those she cares for and loves in a crowd, moving or gesturing towards them. She shows the capacity for empathy, looking intensely at others who cry around her. She loves being around people, experiencing connection with others. When she is tired, she will come and lean against you, letting you know she wants to be cuddled or picked up.

As Ava has grown and continues to develop, she has taught those around her profound, irreplaceable lessons. In my sister's words, 'I imagine so many of the people who suffer from a disability just to partake and be included in the daily interactions you and I have. I see how much she desires to be a part of the activities the other kids her age do. My compassion for people with disabilities has grown hundredfold. I want people to see Ava as a vibrant happy girl and not first see her disabilities.' Ava has made us much more sensitive about her distinctive form of absorbed coping, the importance of manipulating spaces to try and establish a common world with her. And this is vital if we are to truly enable her, to promote her capacities, to validate and encourage her unique voice. If we were to simply look at her *without* her impairment, we would remain unresponsive to the unavoidable challenges of her particular embodiment. So much of her future capacity would be unnoticed, untapped, and neglected. If we were to simply look at her *as* her impairment, her individual personality, and the respect it warrants, would be lost. We reduce her to her body. Even as severe as her motor impairments seem, there is a real person there, capable of her own views, her own perceptions and emotions, who can identify those she knows and feels safe with. Perhaps she'll even vote eventually![9] But recognising her as her own person requires an ethical reorientation in ourselves, where we recognise that relationships – that 'village' – situate and contribute to the capacity of individuals like Ava. There was a truly inspiring blog called *Living with Rett Syndrome* that explored one family's experiences with their daughter, Amy. I want to end the book with an apt quotation:

> Finding out what Amy wants, or what she thinks, almost always depends on finding the right question and asking it in the right way, at the right moment. . . . I know that Amy is dependent on me to 'get it', and that must be annoying and frustrating for her. But when I do, it unlocks another little piece of the mystery of her mind and personality. Unlocking her – figuring out who she is, helping her to be her real self, discovering that she is an opinionated individual with limited tolerance for boring books and no truck with anthropomorphism – feels like my life's work. Each little thing helps me to know her better. Isn't being known, fully known, one of the most important things about being human? Being known, being understood – that's what drives connection. That's what it's all about.[10]

[9] Catriona Moore, 'Making Her Mind Up', *Living with Rett Syndrome* [blog], 15 May 2015.
[10] Catriona Moore, 'Knowing', *Living with Rett Syndrome* [blog], 8 May 2014.

Appendix 1

Overview of the Mental Capacity Act 2005 and Inherent Jurisdiction in England and Wales

The Mental Capacity Act 2005 (MCA) in England and Wales, effective 1 October 2007, resulted from the Law Commission's protracted process to reform mental capacity legislation that began in 1989. S. 1 sets out five core principles:

(1) A person must be assumed to have capacity unless it is established that he lacks capacity.
(2) A person is not to be treated as unable to make a decision unless all practicable steps to help him to do so have been taken without success.
(3) A person is not to be treated as unable to make a decision merely because he makes an unwise decision.
(4) An act done, or decision made, under this Act for or on behalf of a person who lacks capacity must be done, or made, in his best interests.
(5) Before the act is done, or the decision is made, regard must be had to whether the purpose for which it is needed can be as effectively achieved in a way that is less restrictive of the person's rights and freedom of action.

The first three principles [s.1 (2)–(4)] seek to protect the decision-making power and right of autonomy of individuals with impairments. Incapacity is determined through two stages: first, the diagnostic threshold stipulates that there must be evidence of a disturbance or impairment of the mind or brain. Importantly, s. 2 states a person's age, appearance, or behaviour cannot determine capacity. In line with the common law, the MCA forwards a functional test in s. 3, based on the ability

(a) to understand the information relevant to the decision,
(b) to retain that information,
(c) to use or weigh that information as part of the process of making the decision, and
(d) to communicate his decision (whether by talking, using sign language or any other means).

Under the MCA, capacity is context- and decision-specific, meaning that individuals may for example have capacity to consent to marriage or decide about their medical treatment but lack capacity with financial matters, or vice versa.

When individuals are found to lack capacity, best interests decisions can be made on their behalf [s. 4]. According to the MCA, best interests assessors are obliged to consider whether the person might regain capacity about the particular decision in the future and must, 'so far as reasonably practicable, permit and encourage the person to participate, or to improve his ability to participate, as fully as possible in any act done for him and any decision affecting him' [S.4 (4)]. They must consult the person's past and present, wishes, feelings, beliefs, and values (particularly past statements made when the individual was capacitous), as well as the views of others (including lasting powers of attorney or deputies [s. 4 (6)–(7)]. Technically, the MCA adopts a participatory, substituted-judgment model of best interests decision-making. There has been some debate as to the extent to which individuals' preferences and values are determinative in the best interests decision, with some judgments [*LB Haringey v FG* [2011] EWHC 3932 (COP); *W v M, S, NHS Primary Care Trust* [2011] EWHC 2443 (Fam)] suggesting that the further away individuals are from borderline capacity, the less weight their views will have. However, Lady Hale's Supreme Court judgement in *Aintree University Hospitals NHS Foundation v James* [2013] UKSC 67 emphasised that the person's participation in the best interests decision is to be maximised, an interpretation that has been affirmed in subsequent judgments (i.e. *Wye Valley NHS Trust v B* (Rev 1) [2015] EWCOP 60).

The Court of Protection was formed specifically to replace an older court of the same name (which focused more narrowly on the property and financial affairs of incapacitated individuals) and makes judgments in accordance with the MCA regarding decisions about treatment, care, and personal welfare, along with property and financial affairs. The Court also has the power to appoint or revoke deputies as well as enduring and lasting powers of attorney. The MCA and the Court of Protection work in conjunction with other areas of the common law, and indeed, cases outside of the Court of Protection are subject to the same statutory requirements regarding individuals who may lack capacity.

There is, however, some controversy about the blurring boundaries between the MCA and the court's inherent jurisdiction. In *Re F (Mental Patient: Sterilisation) [1990]* 2 AC, the House of Lords held that interventions in decisions affecting incapacitated adults could be exercised under the inherent protective jurisdiction of the court (usually applicable to the wardship of children) so long as the treatment met conditions of 'necessity' and was within the individual's best interests. The pre-MCA case, *Re SA* [2006] 1 FLR further developed inherent jurisdiction in two significant ways: (1) the court's inherent protective jurisdiction was extended to vulnerable adults, who *may or may not* be incapacitated by mental disorder or mental illness, but nonetheless were susceptible to significant harm, exploitation; (2) the judgment involved *pre-emptive* interventions in detrimental environmental conditions where the individual was 'reasonably believed to be, either: (i) under constraint; or (ii) subject to coercion or undue influence; or (iii) for some other reason deprived of the capacity to make the relevant decision, or disabled from making a free choice, or incapacitated or disabled from giving or expressing a real and genuine consent' [para. 77]. A new legal category of 'vulnerable adult' emerged with Munby J's (as he was then) judgment. Prior to the MCA, the court's inherent jurisdiction was exercised in relation to both those with or without capacity.

Inherent jurisdiction has been interpreted as working alongside the MCA. In practice, some jurisdictional bleeding between the two remains inevitable, understandably given their significant overlap, even as both deploy different definitions and substantive and procedural safeguards.[1] Yet difficult questions have emerged recently about the status of inherent jurisdiction vis-à-vis the MCA's statutory provisions. Some worry that inherent jurisdiction tends to undermine or sideline nuances within the MCA.[2] Post-MCA, inherent jurisdiction cases have sanctioned interventions into the lives of capacitous individuals, which appears to fundamentally violate the core principles at the heart of the MCA.[3] This is especially so if the courts subscribe to a narrow interpretation of incapacity as the causative nexus between mental impairment and the inability to decide. Even if one concurs with judgments asserting the survival of the court's protective framework following the MCA, the scope and purpose of inherent jurisdiction has become unclear. Some judgments [*LBL v RYJ and VJ* [2010] EWHC 2665 (COP)] suggest the function of this protective framework is restricted to 'facilitat[ing] the process of unencumbered decision-making by those who they have determined have capacity free of external pressure or physical restraint in making those decisions' [para. 62], whilst others [*NCC v TB & PB* (2014) EWCOP 14] claim that inherent jurisdiction extends beyond injunctive relief to impose a regime against an individual's wishes.

[1] Peter Bartlett and Ralph Sandland, *Mental Heath Law: Policy and Practice*, 4th ed. (Oxford: Oxford University Press, 2014) p. 167.

[2] Ibid., p. 167; 'NCC v PB & TB', 39 Essex Street, available online: http://www.39essex.com/cop_cases/ncc-v-pb-and-tb/ (accessed 31 March 2016).

[3] Indeed, in *XCC v AA & Anor* [2012] EWHC 2183 (COP), Parker J argued that the statutory criteria for best interests decisions under the MCA did not confine decisions under inherent jurisdiction [para. 57]).

BIBLIOGRAPHY

Anderson, Joel and Axel Honneth. 'Autonomy, Vulnerability, Recognition, and Justice'. In John Christman and Joel Anderson. Eds. *Autonomy and the Challenges to Liberalism*. Pp. 127–49.

Agich, George J. *Autonomy and Long-Term Care*. Oxford: Oxford University Press, 1993.

Appelbaum, Paul S. 'Assessment of Patients' Capacities to Consent to Treatment'. *The New England Journal of Medicine* 357:18 (2007): 1834–1840.

Appelbaum, Paul S. and Thomas Grisso. 'The MacArthur Treatment Competence Study. I; Mental Illness and Competence to Consent to Treatment'. *Law and Human Behavior* 19:2 (1995): 105–26.

Appelbaum, Paul S. and Thomas Grisso. 'The MacArthur Treatment Competence Study. II; Measures of Abilities Related to Competence to Consent to Treatment'. *Law and Human Behavior* 19:2 (1995): 127–48.

Atkins, Kim. 'Autonomy and Autonomy Competencies: A Practical and Relational Approach'. *Nursing Philosophy* 7 (2006): 205–15.

Bach, Michael and Lana Kerzner. *A New Paradigm for Protecting Autonomy and the Right to Legal Capacity*. Ontario: Law Commission of Ontario, 2010.

Baron, Marcia and Melissa Seymour Fahmy. 'Beneficence and Other Duties of Love in *The Metaphysics of Morals*'. In Thomas E. Hill, Jr. Ed. *The Blackwell Guide to Kant's Ethics*. Pp. 211–28.

Bartlett, Peter. *Blackstone's Guide to the Mental Capacity Act 2005*. 2nd ed. New York: Oxford University Press, 2008.

Bartlett, Peter and Ralph Sandland. *Mental Heath Law: Policy and Practice*. 4th ed. Oxford: Oxford University Press, 2014.

Beauchamp, Tom L. *Standing on Principles: Collected Essays*. New York: Oxford University Press, 2010.

Beauchamp, Tom L. and James F. Childress. *Principles of Biomedical Ethics*, 7th ed. New York: Oxford University Press, 2012.

Benaroyo, Lazare and Guy Widdershoven. 'Competence in Mental Health Care: A Hermeneutic Approach'. *Health Care Analysis* 12:4 (2004): 295–306.

Benson, Paul. 'Freedom and Value'. *Journal of Philosophy* 84 (1987): 465–86.

Benson, Paul. 'Autonomy and Oppressive Socialization'. *Social Theory and Practice* 17 (1991): 385–408.

Benson, Paul. 'Free Agency and Self-Worth'. *Journal of Philosophy* 91 (1994): 650–68.

Benson, Paul. 'Feeling Crazy; Self-Worth and the Social Character of Responsibility'. In Catriona Mackenzie and Natalie Stoljar. Eds. *Relational Autonomy; Feminist Perspectives on Autonomy, Agency, and the Social Self.* Pp. 72–93.

Benson, Paul. 'Feminist Intuitions and the Normative Substance of Autonomy'. In J. S. Taylor. Ed. *Personal Autonomy: New Essays on Personal Autonomy and its Role in Contemporary Moral Philosophy.* Pp. 133–7.

Bermúdez, José L. 'Normativity and Rationality in Delusional Psychiatric Disorders'. *Mind & Language* 16:5 (2001): 457–93.

Bogden, Robert and Steven J. Taylor. 'Relationships with Severely Disabled People: The Social Construction of Humanness'. *Social Problems* 36 (1989): 135–48.

Brandom, Robert B. *Making It Explicit; Reasoning, Representing, and Discursive Commitment.* Cambridge, MA: Harvard University Press, 1994.

Brandom, Robert B. 'Review: Knowledge and the Social Articulation of the Space of Reasons'. *Philosophy and Phenomenological Research* 55:4 (1995): 895–908.

Brandom, Robert B. *Articulating Reasons; An Introduction to Inferentialism.* Cambridge, MA: Harvard University Press, 2000.

Brandom, Robert B. 'Reply to Charles Taylor's "Language Not Mysterious?"' In Bernhard Weiss and Jeremy Wanderer, Eds. *Reading Brandom on Making It Explicit.* Pp. 301–4.

Brandt, Richard. *A Theory of the Good and Right.* Oxford: Clarendon, 1979.

Brownlee, K. and A. Cureton. Eds. *Disability and Disadvantage.* Oxford: Oxford University Press, 2009.

Buchanan, Allen E. and Dan W. Brock. *Deciding for Others: The Ethics of Surrogate Decision Making.* Cambridge: Cambridge University Press, 1990.

Buss, S. and L. Overton. Eds. *Contours of Agency: Essays on Themes from Harry Frankfurt.* Cambridge, MA: MIT, 2002.

Cave, Emma. 'Determining Capacity to Make Medical Treatment Decisions: Problems Implementing the Mental Capacity Act 2005'. *Statute Law Review* 36:1 (2015): 86–106.

Charland, Louis C. 'Is Mr. Spock Mentally Competent? Competence to Consent and Emotion'. *Philosophy, Psychiatry, and Psychology* 5:1 (1998): 67–81.

Charles J. 'New Guidance issued on Facilitating Participation of "'P'" and Vulnerable Persons in Court of Protection Proceedings'. November 2016. Available at: http://www.familylaw.co.uk/system/froala_assets/documents/1245/Practice_Guidance_Vulnerable_Persons.pdf Accessed 15 November 2016.

Christman, John. 'Autonomy and Personal History'. *Canadian Journal of Philosophy* 21 (1991): 1–24.

Christman, John. 'Relational Autonomy, Liberal Individualism, and the Social Constitution of Selves'. *Philosophical Studies* 117 (2004): 143–63.

Christman, John. *The Politics of Persons; Individual Autonomy and Socio-historical Selves.* Cambridge: Cambridge University Press, 2009.

Christman, John and Joel Anderson. Eds. *Autonomy and the Challenges to Liberalism.* Cambridge: Cambridge University Press, 2005.

Craigie, Jillian. 'Competence, Practical Rationality and What a Patient Values'. *Bioethics* 25:6 (2011): 326–33.

Craigie, Jillian and Alicia Coram. 'Irrationality, Mental Capacities, and Neuroscience'. In Nicole A. Vincent. Ed. *Neuroscience and Legal Responsibility.* Pp. 85–109.

Darwall, Stephen L. 'Internalism and Agency'. *Philosophical Perspectives* 6, Ethics (1992): 155–74.

Darwall, Stephen. *Welfare and Rational Care.* Princeton: Princeton University Press, 2002.

Darwall, Stephen L. *The Second Person Standpoint: Morality, Respect, and Accountability.* Cambridge, MA: Harvard University Press, 2006.

Davidson, Donald. *Inquiries into Truth and Interpretation.* Oxford: Clarendon, 1984.

Davidson, Donald. *Essays on Actions and Events.* Oxford: Oxford University Press, 2001.

Dean, Richard. *The Value of Humanity in Kant's Moral Theory.* Oxford: Clarendon, 2006.

Dempsey, Michelle Madden. *Prosecuting Domestic Violence; A Philosophical Analysis.* Oxford: Oxford University Press, 2009.

Department of Constitutional Affairs. *Mental Capacity Act 2005 Code of Practice.* London: TSO, 2007.

Dhanda, A. 'Constructing a New Human Rights Lexicon: Convention on the Rights of Persons with Disabilities'. *Sur – Revista Internacional de Direitos Humanos* 5 (2008): 42–59.

Dodds, Susan. 'Dependence, Care, and Vulnerability'. In Catriona Mackenzie, Wendy Rogers, and Susan Dodds. Eds. *Vulnerability: New Essays in Ethics and Feminist Philosophy.* Pp. 181–202.

Donchin, Anne. 'Understanding Autonomy Relationally: Toward a Reconfiguration of Bioethical Principles'. *Journal of Medicine and Philosophy* 26 (2001): 365–86.

Donnelly, Mary. 'Capacity Assessment under the Mental Capacity Act 2005: Delivering on the Functional Approach?' *Legal Studies* 29:3 (2009): 464–91.

Donnelly, Mary. *Healthcare Decision-Making and the Law: Autonomy, Capacity and the Limits of Liberalism.* Cambridge: Cambridge University Press, 2010.

Dresser, Rebecca. 'Dworkin on Dementia: Elegant Theory, Questionable Policy'. *Hastings Center Report* 25: 6 (1995): 32–8.

Dreyfus, Hubert L. *Skillful Coping; Essays on the Phenomenology of Everyday Perception and Action.* Oxford: Oxford University Press, 2014.

Dreyfus, Hubert L. and Charles Taylor. *Retrieving Realism.* Cambridge, MA: Harvard University Press, 2015.

Dunn, Michael C. 'When Are Adult Safeguarding Interventions Justified?' In J. Wallbank and J. Herring. Eds. *Vulnerabilities, Care and Family Law.* Pp. 234–53.

Dupré, Catherine. 'Unlocking Human Dignity: Towards a Theory for the 21st Century'. *European Human Rights Law Review* 2 (2009): 190–205.

Dworkin, Gerald. 'Autonomy and Behavior Control'. *Hastings Center Report* 6 (1976): 23–8.

Dworkin, Gerald. *The Theory and Practice of Autonomy.* Cambridge: Cambridge University Press, 1988.

Dworkin, Ronald. 'Natural Law Revisited'. *University of Florida Law Review* 34:2 (1981–82): 165–88.

Dworkin, Ronald. *Life's Dominion: An Argument about Abortion, Euthanasia, and Individual Freedom.* New York: Alfred A. Knopf, 1993.

Dworkin, Ronald. *Taking Rights Seriously.* London: Duckworth, 1996.

Dworkin, Ronald. *Law's Empire.* Oxford: Hart, 1998.

Dworkin, Ronald. 'Rights as Trumps'. In Aileen Kavanagh and John Oberdiek. Eds. *Arguing about the Law.* Pp. 335–44.

Eldergill, Anselm. 'Compassion and the Law: A Judicial Perspective'. *Elder Law Journal* [2015]: 268–78.

Ells, Carolyn. 'Lessons about Autonomy from the Experience of Disability'. *Social Theory and Practice* 27 (2001): 599–615.

Equality and Human Rights Commission. *Human Rights Review 2012; How Fair Is Britain? An Assessment of How Well Public Authorities Protect Human Rights.* EHRC, 2012.

Erler, Alexandre and Tony Hope. 'Mental Disorder and the Concept of Authenticity'. *Philosophy, Psychiatry, and Psychology* 21:3 (2014): 219–32.

Fischer, John Martin and Mark Ravizza, S.J. *Responsibility and Control; A Theory of Moral Responsibility.* Cambridge: Cambridge University Press, 1998.

Fletcher, G. P. 'Human Dignity as a Constitutional Value'. *University of Western Ontario Law Review* 22 (1984): 171–82.

Frankfurt, Harry. 'Freedom of the Will and the Concept of a Person'. *Journal of Philosophy* 68 (1971): 5–20.

Frankfurt, Harry. *The Importance of What We Care About.* Cambridge: Cambridge University Press, 1998.

Fricker, Miranda. *Epistemic Injustice: Power and the Ethics of Knowing.* Oxford: Oxford University Press, 2007.

Friedman, Marilyn. 'Autonomy and the Split-Level Self'. *Southern Journal of Philosophy* 24 (1986): 19–35.

Gadamer, Hans-Georg. *Truth and Method*. 2nd ed. Trans. Joel Winsheimer and Donald G. Marshall. London: Continuum, 2004.

Goldberg, Sanford. 'Comments on Miranda Fricker's *Epistemic Justice*'. *Episteme* 7 (2010): 138–50.

Gooding, Piers. 'Supported Decision-Making: A Rights-Based Disability Concept and Its Implications for Mental Health Law'. *Psychiatry, Psychology and Law* 20 (2014): 431–51.

Grisso, Thomas, Paul S. Applebaum, and Carolyn Hill-Fotouhi. 'The MacCat-T: A Clinical Tool to Assess Patients' Capacities to Make Treatment Decisions'. *Psychiatric Services* 48:11 (1997): 1415–19.

Grondin, Jean. *The Philosophy of Gadamer*. Trans. Kathryn Plant. Chesham: Acumen, 2003.

Gunn, M. J., et al. 'Decision-Making Capacity'. *Medical Law Review* 7 (1999): 269–306.

Hale, Brenda. 'Dignity'. *Journal of Social Welfare and Family Law* 31 (2009): 101–8.

Harman, Gilbert. *Reasoning, Meaning and Mind*. Oxford: Oxford University Press, 1999.

Harpur, Paul David. 'Embracing the New Disability Rights Paradigm: The Importance of the Convention on the Rights of Persons with Disabilities'. *Disability and Society* 27:1 (2012): 1–14.

Held, Virginia. *The Ethics of Care: Personal, Political, and Global*. Oxford: Oxford University Press, 2005.

Herman, Barbara. *The Practice of Moral Judgment*. Cambridge, MA: Harvard University Press, 1996.

Herring, Jonathan. *Caring and the Law*. Oxford: Hart, 2013.

Hill, Jr. Thomas E. *Autonomy and Self-Respect*. Cambridge: Cambridge University Press, 1991.

Hill, Jr. Thomas E., Ed. *The Blackwell Guide to Kant's Ethics*. Chichester: Wiley-Blackwell, 2009,

Honneth, Axel. *The Struggle for Recognition; The Moral Grammar of Social Conflicts*. Trans. Joel Anderson. Cambridge: Polity, 1995.

Hope, Tony, Jacinta Tan, Anne Stewart, and Ray Fitzpatrick. 'Anorexia Nervosa and the Language of Authenticity'. *Hastings Center Report* 41:6 (2011): 19–29.

Hope, Tony, Jacinta Tan, Anne Stewart, and John McMillan. 'Agency, Ambivalence and Authenticity: The Many Ways in Which Anorexia Nervosa Can Affect Autonomy'. *International Journal of Law in Context* 9:1 (2013): 20–36.

Hotopf, Matthew. 'The Assessment of Mental Capacity'. *Clinical Medicine* 5:6 (2005): 580–4.

Jaworska, Agnieszka. 'Respecting the Margins of Agency: Alzheimer's Patients and the Capacity to Value'. *Philosophy and Public Affairs* 28:2 (1999): 105–38.

Kant, Immanuel. *Groundwork of the Metaphysics of Morals*. Trans. H. J. Paton. London: Routledge, 1948.

Kant, Immanuel. *The Metaphysics of Morals*. Trans. Mary Gregor. Cambridge: Cambridge University Press, 1996.

Kavanagh, Aileen and John Oberdiek. Eds. *Arguing about the Law*. Abingdon: Routledge, 2009.

Kayess, R. and P. French. 'Out of Darkness into Light? Introducing the Convention on the Rights of Persons with Disabilities'. *Human Rights Law Review* 8:1 (2008): 1–34.

Killmister, Suzy. 'Dignity – Not Such a Useless Concept'. *Journal of Medical Ethics* 36 (2010): 160–4.

Kittay, Eva Feder. *Love's Labor: Essays on Women, Equality, and Dependency*. New York: Routledge, 1999.

Kittay, Eva Feder. 'At the Margins of Personhood'. *Ethics* 116 (2005): 100–31.

Kittay, Eva Feder and Licia Carlson. Eds. *Cognitive Disability and Its Challenge to Moral Philosophy*. Chichester: Wiley-Blackwell, 2010.

Kong, Camillia. 'The Normative Source of Kantian Hypothetical Imperatives'. *International Journal of Philosophical Studies* 20 (2012): 661–90.

Kong, Camillia. 'Beyond the Balancing Scales: The Importance of Prejudice and Dialogue in *A Local Authority v E and Others*'. *Child and Family Law Quarterly* 26:2 (2014): 216–36.

Kong, Camillia. 'The Space between Second-Personal Respect and Rational Care in Theory and Mental Health Law'. *Law and Philosophy* 34 (2015): 433–67.

Kong, Camillia. 'The Convention for the Rights of Persons with Disabilities and Article 12: Prospective Feminist Lessons against the "Will and Preferences" Paradigm'. *Laws* 4 (2015): 709–28.

Kupfner, Joseph. 'The Moral Perspective of Humility'. *Pacific Philosophical Quarterly* 84 (2003): 249–269.

Lacey, Nicola. *Unspeakable Subjects; Feminist Essays in Legal and Social Theory*. Oxford: Hart, 1998.

The Law Commission. *Mentally Incapacitated Adults and Decision-Making: Medical Treatment and Research*. Law Commission Consultation Paper No. 129, 1993.

The Law Commission. *Mental Incapacity; Item 9 of the Fourth Programme of Law Reform: Mentally Incapacitated Adults*. Law Commission Consultation Paper No. 231, 1995.

Levinas, Emmanuel. *Entre Nous*. Trans. Michael Smith and Barbara Harshav. London: Continuum, 2006.

Lukes, Steven. *Power: A Radical View*. 2nd ed. Basingstoke and New York: Palgrave Macmillan, 2005.

Mackenzie, Catriona. 'The Importance of Relational Autonomy and Capabilities for an Ethics of Vulnerability'. In Catriona Mackenzie, Wendy Rogers, and Susan Dodds. Eds. *Vulnerability: New Essays in Ethics and Feminist Philosophy*. Pp. 33–59.

Mackenzie, Catriona and Natalie Stoljar. Eds. *Relational Autonomy; Feminist Perspectives on Autonomy, Agency, and the Social Self.* New York: Oxford University Press, 2000.

Mackenzie, Catriona, Wendy Rogers, and Susan Dodds. Eds. *Vulnerability: New Essays in Ethics and Feminist Philosophy.* Oxford: Oxford University Press, 2014.

Marzola, Enrica, Giovanni Abbate-Daga, Carla Gamaglia, Federico Amianto, and Secondo Fassino. 'A Qualitative Investigation into Anorexia Nervosa: The Inner Perspective'. *Cogent Psychology* 2:1 (2015).

McCrudden, Christopher. 'Human Dignity and Judicial Interpretation of Human Rights'. *European Journal of International Law* 19 (2008): 655–724.

McDowell, John. 'Knowledge and the Internal'. *Philosophy and Phenomenological Research* 55:4 (1995): 877–93.

McMahan, Jeff. 'Cognitive Disability, Misfortune, and Justice'. *Philosophy and Public Affairs* 25 (1996): 3–35.

McMahan, Jeff. *The Ethics of Killing: Problems at the Margins of Life.* New York: Oxford University Press, 2003.

McMahan, Jeff. 'Radical Cognitive Limitation'. In K. Brownlee and A. Cureton. Eds. *Disability and Disadvantage.* Pp. 240–259.

Mencap. *Death by Indifference: Following Up the Treat Me Right! Report.* London: Mencap, 2007.

Mencap. *Death by Indifference: 74 Deaths and Counting; A Progress Report 5 Years On.* London: Mencap, 2012.

Meyers, Diana T. *Self, Society, and Personal Choice.* New York: Columbia University Press, 1989.

Meyers, Diana T. 'Intersectional Identity and the Authentic Self?: Opposites Attract!' In Catriona Mackenzie and Natalie Stoljar. Eds. *Relational Autonomy; Feminist Perspectives on Autonomy, Agency, and the Social Self.* Pp. 151–80.

Meyers, Diana Tietjens. 'Book Review of Oshana's *Personal Autonomy in Society*'. *Hypatia* 23 (2008): 202–6.

Meynen, Gerben. 'Depression, Possibilities, and Competence: A Phenomenological Perspective'. *Theoretical Medicine and Bioethics* 32:3 (2011): 181–91.

Meynen, Gerben and Guy Widdershoven. 'Competence in Health Care: An Abilities-based versus a Pathology-based Approach'. *Clinical Ethics* 7 (2012): 39–44.

Mill, J. S. *On Liberty.* London: Penguin, 1974.

Montgomery, Jonathan. 'Health Care in Multi-Faith Society'. In John Murphy. Ed. *Ethnic Minorities, Their Families and the Law.* Pp. 161–79.

Moore, Catriona. *Living with Rett Syndrome* [blog].

Murphy, John. Ed. *Ethnic Minorities, Their Families and the Law.* Oxford: Hart, 2000.

Nakazawa, Donna Jackson. *Childhood Disrupted: How Your Biography Becomes Your Biology and How You Can Heal.* New York: Atria, 2015.

Naoki, Higashida. *The Reason I Jump.* Trans. K. A. Yoshida and David Mitchell. London: Sceptre, 2007.

Nedelsky, Jennifer. *Law's Relations: A Relational Theory of Self, Autonomy, and Law.* New York and Oxford: Oxford University Press, 2013.

Nelson, Hilde Lindemann. *Damaged Identities; Narrative Repair.* Ithaca: Cornell University Press, 2001.

Nussbaum, Martha C. *Upheavals of Thought.* Cambridge: Cambridge University Press, 2003.

Nussbaum, Martha C. *Frontiers of Justice: Disability, Nationality, Species Membership.* Cambridge, MA: Belknap, 2007.

O'Neill, Onora. 'Between Consenting Adults'. *Philosophy and Public Affairs* 14 (1985): 252–77.

O'Neill, Onora. *Constructions of Reason: Explorations of Kant's Practical Philosophy.* Cambridge: Cambridge University Press, 1990.

O'Neill, Onora. *Autonomy and Trust in Bioethics.* Cambridge: Cambridge University Press, 2002.

Oshana, Marina. 'Personal Autonomy and Society'. *Journal of Social Philosophy* 29 (1998): 81–102.

Oshana, Marina. 'The Autonomy Bogeyman'. *Journal of Value Inquiry* 35 (2001): 209–26.

Oshana, Marina. 'The Misguided Marriage of Responsibility and Autonomy'. *Journal of Ethics* 6 (2002): 261–80.

Oshana, Marina. 'How Much Should We Value Autonomy?' *Social Philosophy and Policy* 20 (2003): 99–126.

Oshana, Marina. *Personal Autonomy in Society.* Aldershot: Ashgate, 2006.

Press Association. 'More Parents in England Prosecuted for Taking Children out of School'. *The Guardian.* 12 August 2015.

Rawls, John. *A Theory of Justice.* Revised ed. Oxford: Oxford University Press, 1999.

Rosen, Michael. *Dignity: Its History and Meaning.* Cambridge, MA: Harvard University Press, 2012.

Sachs, David. 'How to Distinguish Self-Respect from Self-Esteem'. *Philosophy and Public Affairs* 10:4 (1981): 346–60.

Scanlon, T. M. *What We Owe to Each Other.* Cambridge, MA: Harvard University Press, 2000.

Scanlon, T. M. *Moral Dimensions.* Cambridge, MA: Harvard University Press, 2010.

Schoeman, Ferdinand David. Ed. *Responsibility, Character, and the Emotions: New Essays in Moral Psychology.* Cambridge: Cambridge University Press, 1987.

Sensen, Oliver. 'Kant's Conception of Human Dignity'. *Kant-Studien* 100 (2009): 309–31.

Shakespeare, Tom. *Disability Right and Wrongs*. London: Routledge, 2006.

Siebrasse, Norman. '*Malette v. Shulman*: The Requirement of Consent in Medical Emergencies'. *McGill Law Journal* 34 (1989): 1080–98.

Singer, Peter. 'Speciesism and Moral Status'. In Eva Feder Kittay and Licia Carlson. Eds. *Cognitive Disability and Its Challenge to Moral Philosophy*. Pp. 331–44.

Sneiderman, Barney 'The Shulman Case and the Right to Refuse Treatment', *Humane Medicine* 7:1 (1991): 15–21.

Snow, Nancy. 'Humility'. *Journal of Value Inquiry* 29 (1995): 203–16.

Stavert, Jill. 'UN Convention on the Rights of Persons with Disabilities: Possible Implications for Scotland for Persons with Mental Disorders'. *Scottish Human Rights Journal* 47 (2009).

Stump, Eleonore. 'Control and Causal Determinism'. In S. Buss and L. Overton. Eds. *Contours of Agency: Essays on Themes From Harry Frankfurt*. Pp. 33–60.

Tan, Jacinta O. A., Tony Hope, Anne Stewart, and Raymond Fitzpatrick. 'Competence to Make Treatment Decisions in Anorexia Nervosa: Thinking Processes and Values'. *Philosophy, Psychiatry, and Psychology* 13:4 (2006): 267–82.

Tan, Jacinta, Tony Hope, and Anne Stewart. 'Competence to Refuse Treatment in Anorexia Nervosa'. *International Journal of Law and Psychiatry* 26 (2003): 697–707.

Taylor, Charles. *Philosophy and the Human Sciences: Philosophical Papers 2*. Cambridge: Cambridge University Press, 1985.

Taylor, Charles. *Sources of the Self: The Making of Modern Identity*. Cambridge, MA: Harvard University Press, 1989.

Taylor, Charles. *Philosophical Arguments*. Cambridge, MA: Harvard University Press, 1995.

Taylor, Charles. *Dilemmas and Connections*. Cambridge, MA: Harvard University Press, 2011.

Taylor, J. S. Ed. *Personal Autonomy: New Essays on Personal Autonomy and Its Role in Contemporary Moral Philosophy*. Cambridge: Cambridge University Press, 2005.

Thalberg, Irving. 'Hierarchical Analyses of Unfree Action'. *Canadian Journal of Philosophy* 8 (1978): 211–26.

Thomas, Carol. 'Rescuing a Social Relational Understanding of Disability'. *Scandinavian Journal of Disability Research* 6 (2004): 22–36.

Todes, Samuel. *Body and World*. Cambridge, MA: MIT, 2001.

United Nations Committee on the Rights of Persons with Disabilities. General Comment No. 1 – Article 12: Equal Recognition Before the Law UN Doc. No. CRPD/C/GC/1, adopted at the 11th Session. April 2011.

van de Ven, Leontine, Marcel Post, Luc de Witte, and Wim van den Heuvel. 'It Takes Two to Tango: The Integration of People with Disabilities into Society'. *Disability & Society* 20 (2005): 311–29.

Velleman, J. David. 'The Possibility of Practical Reason'. *Ethics* 106:4 (1996): 694–726.

Vincent, Nicole A. Ed. *Neuroscience and Legal Responsibility*. Oxford: Oxford University Press, 2013.

Wallbank, J. and J. Herring. Eds. *Vulnerabilities, Care and Family Law*. London: Routledge, 2013.

Watson, Gary. 'Free Agency'. *Journal of Philosophy* 72 (1975): 205–20.

Weiss, Bernhard and Jeremy Wanderer. Eds. *Reading Brandom on Making It Explicit*. Abingdon: Routledge, 2010.

Westlund, Andrea C. 'Rethinking Relational Autonomy'. *Hypatia* 24 (2009): 26–49.

Williams, Bernard. *Moral Luck; Philosophical Papers 1973–1980*. Cambridge: Cambridge University Press, 1981.

Wolf, Susan. 'Sanity and the Metaphysics of Responsibility'. In Ferdinand David Schoeman. Ed. *Responsibility, Character, and the Emotions: New Essays in Moral Psychology*. Pp. 46–62.

Wolf, Susan. *Freedom within Reason*. New York and Oxford: Oxford University Press, 1990.

INDEX

Books in the series

Marcus Radetzki, Marian Radetzki, Niklas Juth
Genes and Insurance: Ethical, Legal and Economic Issues

Ruth Macklin
Double Standards in Medical Research in Developing Countries

Donna Dickenson
Property in the Body: Feminist Perspectives

Matti Häyry, Ruth Chadwick, Vilhjálmur Árnason, Gardar Árnason
The Ethics and Governance of Human Genetic Databases: European Perspectives

Ken Mason
The Troubled Pregnancy: Legal Wrongs and Rights in Reproduction

Daniel Sperling
Posthumous Interests: Legal and Ethical Perspectives

Keith Syrett
Law, Legitimacy and the Rationing of Health Care

Alastair Maclean
Autonomy, Informed Consent and the Law: A Relational Change

Heather Widdows, Caroline Mullen
The Governance of Genetic Information: Who Decides?

David Price
Human Tissue in Transplantation and Research

Matti Häyry
Rationality and the Genetic Challenge: Making People Better?

Mary Donnelly
Healthcare Decision-Making and the Law: Autonomy, Capacity and the Limits of Liberalism

Anne-Maree Farrell, David Price and Muireann Quigley
Organ Shortage: Ethics, Law and Pragmatism

Sara Fovargue
Xenotransplantation and Risk: Regulating a Developing Biotechnology

John Coggon
What Makes Health Public?: A Critical Evaluation of Moral, Legal, and Political Claims in Public Health

Mark Taylor
Genetic Data and the Law: A Critical Perspective on Privacy Protection

Anne-Maree Farrell
The Politics of Blood: Ethics, Innovation and the Regulation of Risk

Stephen Smith
End-of-Life Decisions in Medical Care: Principles and Policies for Regulating the Dying Process

Michael Parker
Ethical Problems and Genetics Practice

William W. Lowrance
Privacy, Confidentiality, and Health Research

Kerry Lynn Macintosh
Human Cloning: Four Fallacies and Their Legal Consequence

Heather Widdows
The Connected Self: The Ethics and Governance of the Genetic Individual

Amel Alghrani, Rebecca Bennett and Suzanne Ost
Bioethics, Medicine and the Criminal Law Volume I: The Criminal Law and Bioethical Conflict: Walking the Tightrope

Danielle Griffiths and Andrew Sanders
Bioethics, Medicine and the Criminal Law Volume II: Medicine, Crime and Society

Margaret Brazier and Suzanne Ost
Bioethics, Medicine and the Criminal Law Volume III: Medicine and Bioethics in the Theatre of the Criminal Process

Sigrid Sterckx, Kasper Raus and Freddy Mortier
Continuous Sedation at the End of Life: Ethical, Clinical and Legal Perspectives

A. M. Viens, John Coggon, Anthony S. Kessel
Criminal Law, Philosophy and Public Health Practice

Ruth Chadwick, Mairi Levitt and Darren Shickle
The Right to Know and the Right not to Know: Genetic Privacy and Responsibility

Eleanor D. Kinney
The Affordable Care Act and Medicare in Comparative Context

Katri Lõhmus
Caring Autonomy: European Human Rights Law and the Challenge of Individualism

Catherine Stanton and Hannah Quirk
Criminalising Contagion: Legal and Ethical Challenges of Disease Transmission and the Criminal Law

Sharona Hoffman
Electronic Health Records and Medical Big Data: Law and Policy

Barbara Prainsack and Alena Buyx
Solidarity in Biomedicine and Beyond

Camillia Kong
Mental Capacity in Relationship

Oliver Quick
Regulating Patient Safety: The End of Professional Dominance?

Thana C. de Campos
The Global Health Crisis: Ethical Responsibilities

Jonathan Ives, Michael Dunn, Alan Cribb (eds)
Empirical Bioethics: Theoretical and Practical Perspectives

Alan Merry and Warren Brookbanks
Merry and McCall Smith's Errors, Medicine and the Law: Second Edition

Donna Dickenson
Property in the Body: Feminist Perspectives, Second Edition

Printed in the United States
By Bookmasters